AF695939

Light Weight Materials

Light Weight Materials

Processing and Characterization

Edited by

Kaushik Kumar
Bathini Sridhar Babu
J. Paulo Davim

iSTE

WILEY

First published 2021 in Great Britain and the United States by ISTE Ltd and John Wiley & Sons, Inc.

Apart from any fair dealing for the purposes of research or private study, or criticism or review, as permitted under the Copyright, Designs and Patents Act 1988, this publication may only be reproduced, stored or transmitted, in any form or by any means, with the prior permission in writing of the publishers, or in the case of reprographic reproduction in accordance with the terms and licenses issued by the CLA. Enquiries concerning reproduction outside these terms should be sent to the publishers at the undermentioned address:

ISTE Ltd
27-37 St George's Road
London SW19 4EU
UK

www.iste.co.uk

John Wiley & Sons, Inc.
111 River Street
Hoboken, NJ 07030
USA

www.wiley.com

© ISTE Ltd 2021
The rights of Kaushik Kumar, Bathini Sridhar Babu and J. Paulo Davim to be identified as the authors of this work have been asserted by them in accordance with the Copyright, Designs and Patents Act 1988.

Library of Congress Control Number: 2021947485

British Library Cataloguing-in-Publication Data
A CIP record for this book is available from the British Library
ISBN 978-1-78630-797-2

MIX
Paper from responsible sources
FSC® C013604

Contents

Preface

We would like to present the book *Light Weight Materials: Processing and Characterization*. In the automotive industry, the need to reduce vehicle weight has led to extensive research efforts to develop aluminum and magnesium alloys for structural car body parts. In aerospace, the move towards composite airframe structures has led to an increased use of formable titanium alloys. All of the above-mentioned materials can be categorized into a group called "*lightweight materials*". The distinguishing feature of lightweight materials is their low densities, ranging from as low as 0.80 g/cm^3 for unfilled polymers to as high as 4.5 g/cm^3 for titanium. Although the density of titanium is higher than that of unfilled polymers, it is significantly lighter than metals: alloy steel (7.86 g/cm^3) and superalloys (7.8–9.4 g/cm^3). In a nutshell, lightweight materials exhibit a wide range of properties and therefore offer a wide range of applications.

This book primarily aims to provide researchers and students with an overview of the recent advancements in the processing, manufacturing and characterization of lightweight materials, which promises increased flexibility in manufacturing in tandem with mass communication, improved productivity and better quality. It has a collection of chapters contributed by eminent researchers who focus on the topics associated with lightweight materials, including the current buzzword *composite materials*. This book provides the recent advancements in the processing, manufacturing and characterization of lightweight materials and hence would be a panacea in all areas of lightweight materials.

This book has two major objectives. Firstly its chapters by eminent researchers in the field enlighten readers about the current status of the subject. Secondly, as the densities vary a lot so do the applications ranging from automobile, aviation to bio-mechatronics; hence, this book would serve as an excellent guideline for people in all of these fields.

The chapters of this book are divided into three parts, namely Part 1: Manufacturing Processing Techniques, Part 2: Characterization and Part 3: Analysis.

Part 1 contains Chapters 1–3, Part 2 contains Chapters 4 and 5 and Part 3 contains Chapters 6–8.

Chapter 1 explains an advanced technique called *additive manufacturing* (AM), which is predominantly known as 3D printing and rapid prototyping. It is an on-demand production without any dedicated apparatus or tooling, which allows breakthrough performance and supreme flexibility in industries. The aerospace industry is the primary user of AM, as it enables it to create complex user-defined part design and fabricate with different lightweight materials without wastage of raw materials, reducing the time and cost of production. This chapter provides in-depth knowledge about its classification and selection process for various applications required by engineering industries, especially in the aerospace industry.

Chapter 2 mainly deals with the manufacturing of polymer gears. Polymer gears are widely used in medical devices upon which human lives depend. In addition, they are useful in other applications such as in the automotive and manufacturing industries. A precise gear of better design and effective manufacturing process decides its long-term application, strength and property. Polymer gears can be fabricated with the same machining process as metal gears, usually milling or hobbing from a blank. However, for lightweight materials, such as polymers, it is preferable to be either fabricated by injection molding or machined from a rod (additive manufacturing). The details of such manufacturing techniques are presented in this chapter.

Chapter 3, the last chapter of Part 1, discusses in detail reinforcing, performance analysis, processing and characterization of various methods of polymer welding, i.e. laser welding, infrared welding, spin welding, stir welding, and vibration welding. This chapter also covers various alloys of aluminum for lightweight applications and the current status of polymer composite applications in industries and future prospects. This chapter highlights the complications related to fusion, heat transfer and joint strength, as well as their solutions with the future prospect of polymer welding empowering polymers to be an absolute substitute for metal, which can be achieved by understanding the concept of dissimilar welding for joining polymer composites with metals and their controlling factors, and by selecting an appropriate welding process for various types of polymers.

Chapter 4, the first chapter of Part 2, provides the reader with an idea of fabrication and a description of the processing techniques of natural-based composites for light body vehicle applications. In doing so, the genetic equation for modeling tool flank wear is developed using experimentally measured flank wear values and genetic programming. Using these results, the genetic model presenting the connection between cutting parameters and tool flank wear is extracted. Then, based on a defined machining performance index and the obtained genetic equation, optimum cutting parameters are determined. This chapter concludes that the proposed modeling and optimization methodology offer the optimum cutting parameters and can be implemented in real industrial applications.

Chapter 5 presents the response surface methodology, an optimization technique, to design a catalytic cracking experiment of plastic waste. The catalyst-to-feedstock ratio, the operating temperature and the reaction time were chosen as an effective parameter of the catalytic cracking process. The characterization of the obtained liquid product was performed using the Fourier transform with infrared (FTIR) spectra, gas chromatography with mass spectrometry (GC/MS) analysis and physico-chemical analysis. This chapter concludes that the developed quadratic model is well fit to the experimental domains and predicts operating conditions that are most suitable for conducting catalytic cracking experiments under recycling techniques of lightweight materials, especially plastics.

Chapter 6, the first chapter of Part 3, discusses laser welding. The uniqueness of this chapter is the way it has dealt with the subject. The finite element analysis was used to select suitable models for the Gaussian beam profile and the application of the Frustum model to conduction mode welding and keyhole laser welds. Temperature and stress analysis was carried out within and around the weld region. This chapter discusses the analytical comparative approximation of different model approaches applicable to the laser weld process, and indicates that the parametric study information will be useful to the engineers of nuclear fabrication applications in finalizing different components.

Chapter 7 elaborates on the effect of formability parameters on tailor-welded blanks of lightweight materials. The product finds its maximum application in the automotive manufacturing industry. It is quite common that different materials with varying cross-sections are used based on the requirements in aerospace and automotive industries. To manage the herculean task of organizing this, researchers have enthusiastically proposed a tailor-made welded blanks (TWB) strategy, and in many automotive industries this technique has been adopted. This chapter suggests testing the formability of tailor-welded blanks with various light alloy sheets used in the aerospace and automotive industries. An overall review of various parameters that affect the formability of tailor-welded blanks is presented in this chapter, so that other investigators can rely on the same for more critical observations in this field.

Chapter 8, the last chapter of this section, presents the various ways of optimizing a vehicle body, such as shape optimization for aerodynamics and aesthetics, and weight of materials to be used for fuel efficiency, material conservation, recyclability and others. This chapter considers a product called "B-pillar", one of the critical structural support members of sedan cars. They have replaced the existing material with a composite, mainly to overcome the stress developed due to the system as it is a structural member and to safeguard the occupant in the case of a side crash. Different mechanical properties such as tensile, compression and bending strength, as well as water absorption, were measured. The model of the sedan car B-pillar panel developed was analyzed for impact and crush simulation. It concluded that a composite can be used for the outer panel of B-pillar, which results in reduced vehicle weight and fuel consumption and increased energy absorption.

First and foremost, we would like to thank God. It was your blessing that provided us with the strength to believe in passion and hard work and to pursue our dreams. We thank our families for having the patience with us for taking yet another challenge that decreased the amount of time we could spend with them. They are our inspiration and motivation. We would like to thank our parents and grandparents for allowing us to follow our ambitions. We would like to thank all the contributing authors, as they are the pillars of this structure. We would also like to thank them for believing in us. We would like to thank all of our colleagues and friends in different parts of the world for sharing their ideas helping us to shape our thoughts. We will be satisfied with our efforts when the professionals concerned with all the fields related to lightweight materials are benefitted.

We owe a huge thanks to all of our technical reviewers, Editorial Advisory Board members, Book Development Editor and the team at ISTE Ltd for their availability to work on this huge project. All of their efforts helped us to complete this book, and we could not have done it without them.

Last, but definitely not least, we would like to thank all of the individuals who have taken time out and helped us during the process of editing this book. Without their support and encouragement, we would have probably given up the project.

Kaushik KUMAR
Bathini SRIDHAR BABU
J. Paulo DAVIM
September 2020

Part 1

Manufacturing Processing Techniques

fixtures, prototypes and support structures in a slow and cost-effective process.

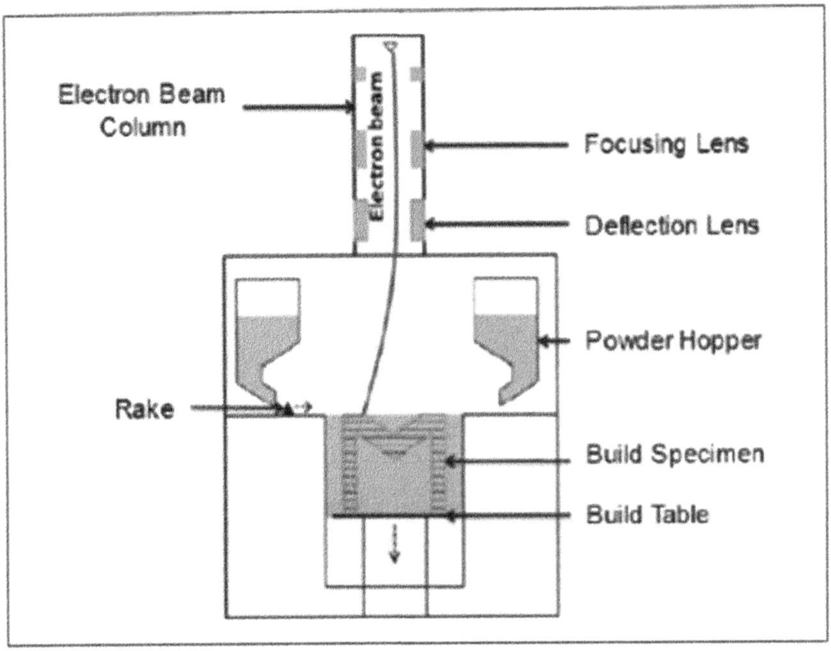

Figure 1.4. *Electron beam melting (Mandil, 2016). For a color version of this figure, see www.iste.co.uk/kumar/materials.zip*

1.3.3. *Selective laser melting*

Selective laser melting (SLM) uses a high-power density laser to melt the raw material and fuse it together with metallic powders, which is mainly used for low-volume materials. In this technique, various materials such as glass, ceramics and plastics are used. The laser beam will heat the particles at appropriate positions on a bed of metallic powder until it is completely smelted. The AM machine will consecutively increase the melted layers over the metal bed until it reaches the expected design (Jürgen 2017). Its common applications can be found in aerospace industries, automobile industries and medical industry to overcome the demand for human organs.

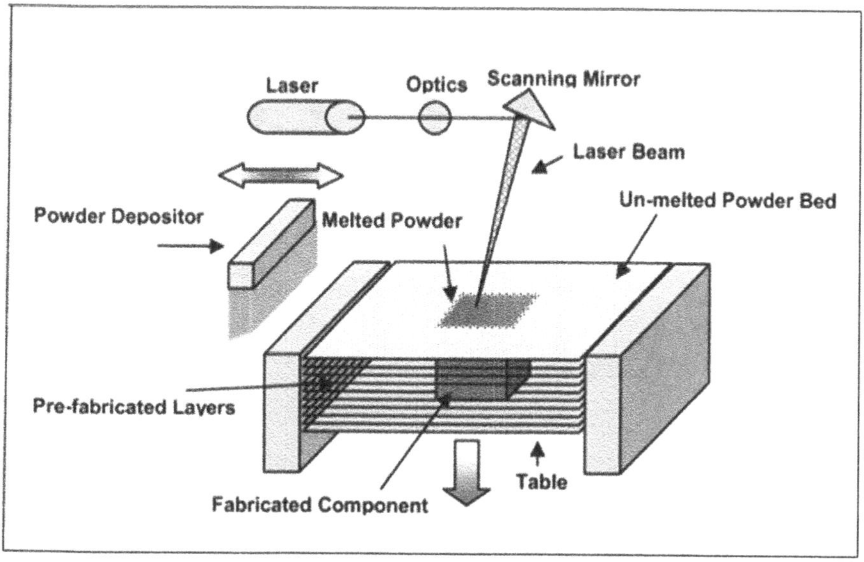

Figure 1.5. *Selective laser melting (Mumtaz 2008). For a color version of this figure, see www.iste.co.uk/kumar/materials.zip*

1.3.4. *Direct metal laser sintering*

Direct metal laser sintering (DMLS) is similar to selective laser melting as the intent is to use a laser with high power density to create the geometry that is best suited for manufacturing metals and metal alloys. DMLS uses a variety of alloys for allowing efficient hardware to design complex geometries. The functional prototypes made using this method have greater strength and ultimate durability. DMLS is most suitable for fabricating complex oil and inert gas components. Metal alloys such as aluminum alloys, stainless steel and niche alloys are widely used (Rizzuti 2019). DMLS materials are completely dense, highly robust and have greater resistance to corrosion, which can be auxiliary treated with heat, sterilization and coating.

corrosion-resistant properties. When heated, Inconel forms an abundant, unchanging, protective oxide layer, which protects the external surface from further attack. The widespread temperature range of Inconel maintains its strength, which is found to be adaptable to high-temperature components of high thermal valency. In the aeronautical field, a variety of aircraft mechanisms and engines for jets, such as lock panels, exhaust linings and blades for turbine stamps, are made of Inconel 600 alloy. It is an accustomed engineering material for solicitations, providing protection from corrosion and heat. It has admirable mechanical properties such as high strength and better workability (Wei 2019). Inconel-713C has been widely used as a super alloy due to its specialized characteristics consisting of intrinsic capability of casting, stabilizing and higher level of strength, and it also acts as a ductile material at increasing temperatures, which are used for blading jet engines. The additional grouping of Inconel is the Inconel 718 which accounts for an up to 50% reduction of the mass of turbojet engines involved in aircrafts.

1.4.3. *Aluminum*

Metallurgy has a vital role in the enlargement of aviation. Aluminum plays a major role in aircraft construction due to its excellent strength-to-weight ratio and cost ratio. It is especially suitable for aircraft manufacture because it is considered as one-third the weight of steel, allowing an aircraft to import maximum weight or renovate the efficiency of fuel consumption. Aluminum alloys have remarkable quality for use in the fuselage, wing and the erection of commercial airliners, military cargo and aircraft. The fabricated and fashioned aluminum are used in structural components of aircraft for the United States Navy. Aluminum can be used as a crucial propellant for motors of space shuttle as its volumetric energy density is great and very hard to burn. The strength of aluminum enables it to substitute heavy metals without loss of strength along supplementary metals (Aboulkhair 2019). Moreover, load-carrying assemblies can yield the benefit of the strength of aluminum to promote aircraft fabrication more consistent with reduced cost. Aluminum is extremely resistant to corrosion as well as to chemical reactions; hence, it is unambiguously suitable for aircraft operations in extremely corrosive environments. The 6063 aluminum is known for promoting prototypical quality attributes, and it is the most suitable alloy for anodizing solicitations. The recipient choice for aerospace applications is Al-7050, as it preserves its strength and properties in wider segments and provides protection against fractures and corrosion.

1.4.4. *Stainless steel*

Stainless steel or inox steel (steel alloy) comprises 11% of the mass of chromium and an extreme 1.2% of the mass of carbon. It encompasses unpredictable amounts of carbon, silicon, manganese and some additional metallic materials such as nickel and molybdenum that are added to inculcate other expedient properties such as improved formability and resistance to corrosion. Its excellent corrosion-resistant property and tolerance to withstand high temperature in fabricating spacecraft components lead to longer durability of the component. Its high tensile strength and shear modulus make it more suitable for absorbing the impact stresses that landing gear equipment should endure (Martina 2019). The most abundantly recycled stainless steel worldwide is T-304, which can be weldable and machinable, and has better resistance to many chemical corrodents as well as industrial atmospheres and has well-known formability. Its austenitic structure offers excellent ductility and formability, which condenses work-hardening tendency and enhances suitability for deep drawing.

1.5. Aerospace applications of additive manufacturing

Over the last five years, AM has expanded from tooling and prototyping to explicit part production, especially to enhance industrialized fields such as aerospace, architecture, automotive and medical science. AM stands as a global technology to bring rapid developments in designing and modeling. AM technology is economically viable and provides the complete potential to process individualized products with complex geometries. Its flexibility to use any kind of raw material such as both metals and non-metals provides better replacement of materials, thereby reducing the cost of production. AM contributes its foremost capability to serving the aerospace industry, because this industry faced a great challenge to reduce the weight and size of components in order to avoid dangers and promote safety and durability. Designing and manufacturing flight-worthy components from high-performance materials are needed for commercial and military aerospace platforms, which can be achieved by AM. 3D printing (3DP) offers multifaceted, amalgamated parts with extraordinary strength. The involvement of less material and consolidated designs results in the overall reduction in weight, which acts as the most important factor in manufacturing for aerospace (Russell 2019). AM is often used for aerodynamic analysis, which has acute prominence for aerospace. The selective laser melting and binder jetting are used to create strong-featured, plane scale prototypes of aerospace designs to yield exact models that allow design indentation and feasibility. 3DP is generally used to produce structural, low-volume metal supports (with DMSL/SLM), which have intricate lifesaving systems, for the internal barricade of a plane. The extremely appropriate technique of prototyping is also ultimate for aerodynamic test results and analysis, as the stable surface finish factor is able to meet the appropriate final part through the 3DP technology. AM has an extraordinary degree of design freedom that assists in optimizing the parts and incorporating multiple features in a single component (Kamal 2019). AM greatly supports the aerospace industry as it is mainly focused on components with multifarious competence and low volume production. The assembly of composite designs with various sizes and the scope that prevails manufacturing in the required volume are achieved in a single machine through AM.

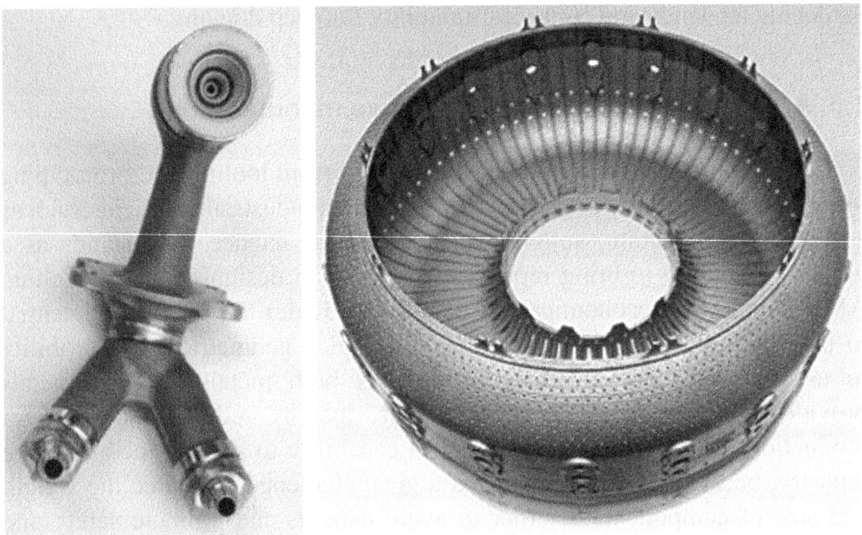

Figure 1.9. *Aerospace components fabricated through AM (Hiemenz 2014)*

1.6. Challenges faced in the aerospace industry

One of the biggest challenges faced in the aerospace industry is the inability to manufacture large and odd-sized components. Another main challenge faced in the field of aerospace is that the strength in the plane of layers is not uniform. The scalability of traditional manufacturing methods is limited, which find difficulty in stocking inventories. Printing multiple materials cannot be achieved simultaneously with AM, thus making it somewhat unsatisfactory. This is a problem that needs to be solved with advances in material science, by expanding the selection of materials and reducing their cost.

1.7. Overcoming aerospace challenges with AM

AM in aerospace gives a good inspection of the parts in a completely automated process. In the manufactured parts in aerospace, an accuracy of about 30–40 microns is offered by AM. Good quality parts along with good tolerance levels of less than 10 microns can be achieved with AM. The flexibility of AM allows the designers to manufacture various parts using different materials, thus increasing its competitiveness. Production and unloading can occur simultaneously with AM technology incorporated in the aerospace industry to overcome the scalability criteria.

1.8. Future work

This chapter explained in detail an appraisal of the methodologies and suitable materials used in AM for advancing the aerospace industry with the performances of topological and material optimization, design integration and part consolidation that promotes profitable quantity and quality products:

– The progression and characteristics of AM technology were studied, along with the impact of material properties and processing operations such as orientation and part performance relevant to specific user-defined part design, which should be developed to enhance the shape and structure of alloyed elements.

– The different new materials with varying mechanical and metallurgical properties should be examined to find a better strength-to-weight ratio than the existing alloys.

designed based on the American Gear Manufacturers Association (AGMA) standard for polymer gears. To satisfy the AGMA standard, gears should have the following properties:

I. high mechanical properties to withstand the force transmitted on them;

II. high dynamic properties to carry fluctuating loads;

III. high surface resistance to overcome pitting failure.

Most importantly, spur gears are widely used for various applications in which power is transmitted between parallel shafts. Spur gears have straight teeth, which are the simplest in design. It is mounted on a parallel shaft to mesh with other gears. It is used to increase or decrease the torque, power or speed of a given object. To improve a spur gear, the gear material should be a polymer. Polymer gears are replacing metal gears in many applications due to their ease of manufacturing, low cost, self-lubrication behavior, noiseless operation, chemical inertness, etc. These gears are useful for transferring power from low to medium transmission, such as in equipment, toys, printers, meters, facsimiles, copy machines, fax and paper mill gears. Polymer materials also have a wide range of components that are used in the design of medical devices for high speed and load applications, in all varieties of gear trains and across a wide range of velocity ratios. The medical device industry requires high quality, smooth and quiet operation of the machine, such as intravenous infusion pump machines, to ensure its proper function without fail as people's lives depend on it. Fine pitch gears are required for this kind of machine. Nowadays, gears are used in numerous medical equipment such as surgery tables, medical diagnostic machines, patient bed gear motor and gear pumps. In an automatic syringe injection pump, the tube is filled with fluid that consists of food or drugs (Jafarzadeh and Farokhi 2016). By means of the movement of gear and the rotation of the roller, food or drugs are given to a patient who is unable to ingest orally. Thus, gears play a big role in controlling the flow of liquid. The gear manufacturing process has an important role in deciding its suitability, affectivity, functionality and quality. Polymer gears are basically prepared from thermoplastic polymers that may be biodegradable or non-biodegradable polymers (Adams 1986). Polybutylene terephthalate (PBT), high-density polyethylene (HDPE), acetal (polyoxymethylene – POM), polyether ether ketone (PEEK), nylon (polyamide) (PA), acrylonitrile butadiene styrene (ABS), polyphenylene sulfide (PPS) and polycarbonate (PC) are classified as non-biodegradable polymers. Biodegradable polymers are polylactide (PLA) and polybutylene succinate (PBS). This chapter deals with the gear manufacturing process, which is explained below.

2.2. Gear manufacturing process

Metal gears are manufactured using different processes such as hobbing, die casting, powder metallurgy, extrusion, rolling, blanking in press tool, wire EDM, etc. Different gear manufacturing processes for metals and non-metals are given in Table 2.1. These manufacturing processes vary from each other depending on the size, production rate and quality of the gear. However, the manufacturing of a polymer spur gear is quite different from that of a metallic spur gear. It may vary due to a low melting point, as well as poor thermal, electrical and mechanical properties. Some of the important manufacturing processes of polymer spur gears are hobbing, injection molding, additive manufacturing/rapid prototyping, etc.

Type of manufacturing processes	Nature of manufacturing processes	Name of manufacturing processes	Type of materials
Traditional manufacturing process	Material subtraction	Hobbing Milling	Ferrous and non-ferrous, polymers
	Material deformation	Extrusion Stamping Forging Hot embossing	Metals
			Polymers
	Material addition	– Injection molding – Die casting – Powder metallurgy – LIGA – 3D printing	Metals, plastics
			Non-ferrous materials
			Metallic and non-metallic materials
			Metallic, alloys and polymers
Non-traditional manufacturing process	Spark erosion	EDM, Wire EDM	Electrical conductive materials
	Chemical ablation	Bio-etching	Metals
	Thermo-electrical process	Laser ablation	Metals, polymers
	Mechanical erosion	Water jet abrasive machining	Metals, non-metals

Table 2.1. *Different manufacturing processes for metal and non-metal gears*

2.2.1. *Gear hobbing machine*

Tool-work arrangements of hobbing are shown in Figure 2.1. A gear milling cutter and the gear blank work together like a pair of worm and worm wheel. This process is the most accurate and highly productive manufacturing process for cutting straight or helical teeth of spur gears and worm wheels. For this purpose, a gear hobbing machine requires some complicated and expensive tools. Spur gears can be easily fabricated by the gear hobbing machine process. It is a complex manufacturing process for the soft machining of gears due to several tooth engagements with the workpiece. Hence, the cutting force of each tooth should be determined using the fly-cutting test procedure (Stark *et al.* 2013).

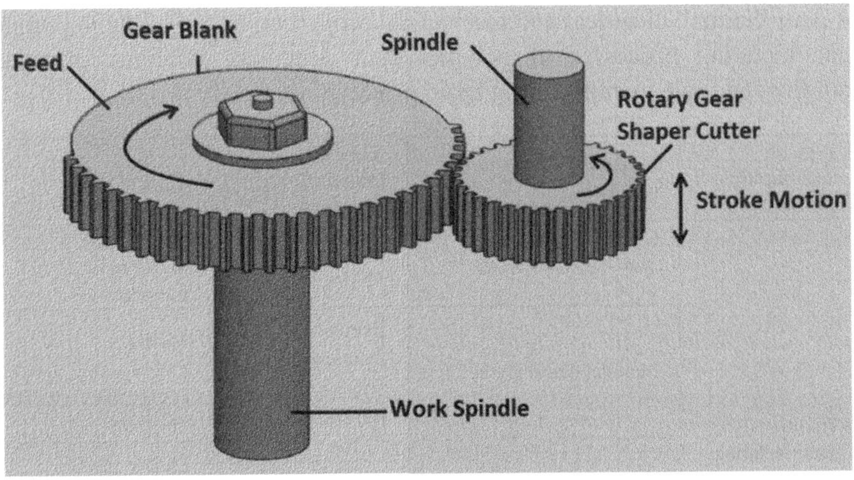

Figure 2.1. *Tool-work arrangements in hobbing*

The test is also used for calculating tooth wear. It has been found that there is no difference between the wear failure mechanism of injection molding and machine cut polymer gears. Before fabricating the gear, thermal deviations and temperature distributions should be determined using the gear hobbing machine process. Hence, the fly-cutting test is modified by IFQ to obtain the real temperature distribution data on the gear tooth. Sudhagar *et al.* (2019) prepared the spur gear using the gear hobbing process, which is made up of madar fiber-reinforced polyester composites, Bauhinia Racemosa fiber-reinforced polyester composites and unsaturated polyester materials, as well as tested their wear behavior and service life. The efficiency of a spur gear is calculated using the following equation (Townsend 1991):

$$E = 100\left[1 - \frac{\mu}{2\cos\alpha}\frac{V_1^2+V_2^2}{V_1+V_2}\right]\% \qquad [2.1]$$

where V_1 and V_2 are the maximum specific sliding velocities of the pinion and the gear, μ is the friction coefficient and α is the operating pressure angle. Here, we present one example of power transmission in an automobile gearbox using plastic gears, which was designed by Kapelevich and McNamara (2005). This gearbox can work at a temperature range of -48°C to +135°C. The plastic gear is used to reduce the cost of the gearbox. The thermal expansion of each component used in this gearbox is different.

Figure 2.2. *An automobile gearbox in which a plastic gear is used as the gear and a metal gear is used as the pinion for power transmission (Kapelevich and McNamara 2005). For a color version of this figure, see www.iste.co.uk/kumar/materials.zip*

2.2.2. *Injection molding*

Plastic injection molding is the most economical when it comes to mass production. Melted plastic is poured into metal dies under pressure, and this method is called "injection molding". Gears obtained from machining operations are useful when a small number of gears are required. However, the molding process is widely used to manufacture gears due to the following reasons:

– no machining operations;

– mass production;

– elimination of preliminary or secondary operations;

– low cost of the injection molding process;

– uniformity of parts.

It is more easy and convenient to make a gear from the injection molding technique than a metal, due to the easy melting of plastic when heated, less shrinkage and high fluidity. Through injection molding, the shape of the gear can be exactly the same as the die used for gear making. However, it is difficult to achieve high-dimensional accuracy while making large gears. In such a case, after injection molding, the gear undergoes a light shaving to acquire the required size and shape of the tooth. The parameters for injection molding are drying temperature, injection pressure, press temperature, mold temperature, waiting time in the mold and total cycle time in seconds. Table 2.1 shows the process parameters for the injection molding of different polymers. These parameters vary from polymer to polymer depending on the physical and chemical properties.

Composite	Injection time (s)	Packing time (s)	Cooling time (s)	Packing pressure (%)	Melt temperature (°C)	Mold temperature (°C)	Injection pressure	References
PA66 + GF50 wt%	1.5	40	120	80	290	10	-	8
	3	30	10	60	220	20	-	
ABS	-	-	-	-	-	55	55	
HDPE	-	-	25	-	-	35	60	9
POM	-	-	-	-	-	60	70	
POM GF+POM	-	-	-	70	176 170	-	-	10

PA: polyamide, GF: glass fiber, ABS: acrylonitrile butadiene styrene

HDPE: high-density polyethylene, POM: polyoxymethylene

Table 2.2. *Process parameters for injection molding for different polymer composites*

The plastic gear obtained from this manufacturing process is used in a wide range of applications ranging from household utensils to automotive industries (Mao 2007). Material selection and proper mold design play an important role in achieving a high-quality gear. While designing a molded plastic gear, a proper gating system, cooling channels, venting, precision fits and tolerances, sufficient coring and ejection capabilities, quality mold surface, finish and proper mold material selection are crucial for designing a good plastic mold. The improper design of the mold can lead to gear failure, with a further cost to modify the running process of production.

Nylon 6/6 spur gears were manufactured using injection molding by Senthilvelan and Gnanamoorthy (2006). The authors reported that molded gear accuracy was significantly influenced by material shrinkage behavior and fiber orientation. They reinforced the short glass fiber in nylon 6/6. Moreover, simulation of gear manufacturing was carried out, which showed a good correlation between the experimental and simulated results. However, the pressure angle correction and module correction methods were widely followed to provide shrinkage compensation in the case of injection-molded polymer gears. However, these methods were based on the postulation that shrinkage would occur towards the center of a molded gear, which is not always possible in real cases (Lee *et al.* 1998). In reinforced polymer spur gears, the performance is decided by fiber orientation and their surface conditions Senthilvelan and Gnanamoorthy (2004a, 2004b). Therefore, Luscher *et al.* (2000) fabricated the polymer gear using injection molding and studied the transmission errors and different geometries by varying gate configurations. They reported the influence of the gate configuration and the pressure on lead vibration and transmission error. There are two types of spur gear: one is symmetric, which is used to transmit power in two directions, and the other is asymmetric, which is used to transmit power in only one direction. There are only a few studies available in the literature regarding asymmetric polymer gears for which experimental investigations have been carried out. In this series, Kumar *et al.* (2009) developed asymmetric polymer gears with a 3 mm module and 18 teeth using the injection molding process at 34 and 20 degree pressure angles. They tested the gears at loads of 1.5, 1.8 and 2.0 N-m at 1,000 RPM for 3 hours. These gears can be driven either in a clockwise or anti-clockwise direction. They reported that the efficiency increased with an increase in the load. Furthermore, Sardar and Bandopadhya (2014) developed and fabricated spur gears from cement-reinforced polypropylene composite (PPC) materials.

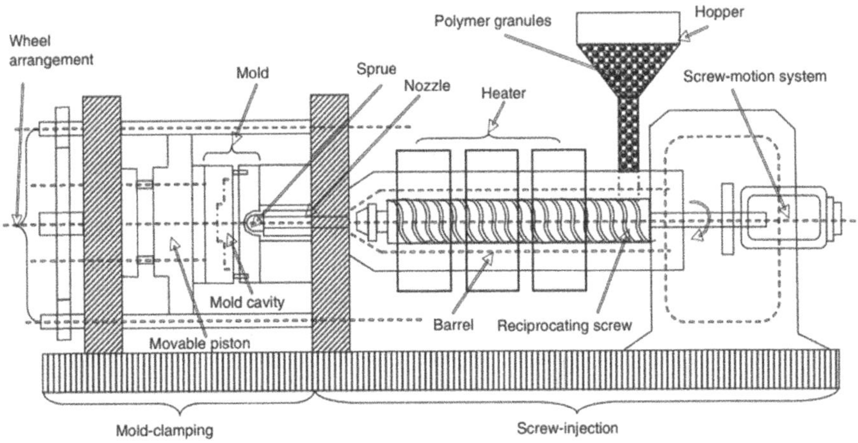

Figure 2.3. *Injection molding setup for fabrication of polymer spur gears (Sardar and Bandopadhya 2014)*

Figure 2.4. *Fabricated composite spur gears: (A) 100% polypropylene (PP) and (B) 10% Portland pozzolana cement (PPC) particle-filled PP (Sardar and Bandopadhya 2014)*

They used injection molding, which is the widely used setup in many cases, as shown in Figure 2.3. They prepared a PPC solution using distilled water in a 3:1 proportion, followed by stirring for 45 minutes in an ultrasonic chamber. Then, the proposed solution was dried at room temperature for 72 hours. It was kept in three heaters maintained at constant temperatures (220, 225 and 230°C). Other parameters studied are the injection shot capacity

(100 g/s), the injection time (8 sec), the cooling time (12 sec) and the mold temperature (28°C). The fabricated spur gears are shown in Figure 2.4.

2.3. Additive manufacturing/rapid prototyping

This technique refers to the direct manufacture of end products using additive fabrication methods. In this method, a model of the product is designed first on a digital platform like CAD, FEM without the use of any tool or mold. It is mainly used to manufacture gear wheels of non-standard (non-involute) tooth profile (Budzik 2011). It not only reduces the time of manufacturing the complex shape but also reduces the cost. In addition, several other benefits of additive manufacturing have motivated gear manufacturers to adopt this technique, such as high quality and good surface integrity of gears, flexibility, minimum waste, energy efficient, customer-ordered production and environmental benefits due to low emissions. To meet the gear fabrication requirements, different methods of additive manufacturing are available for making polymer gears, as shown in Figure 2.5.

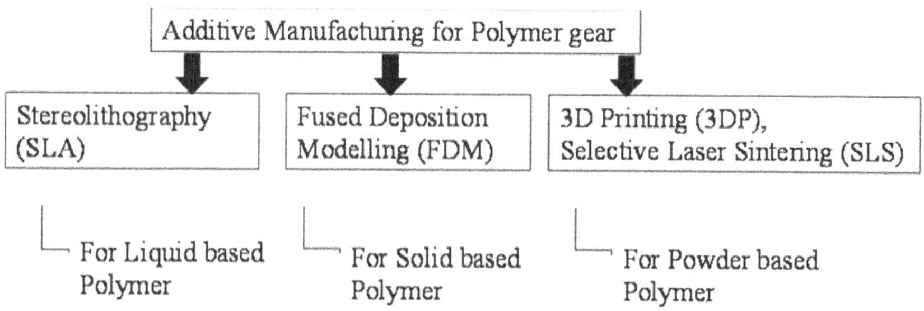

Figure 2.5. *Types of additive manufacturing of polymer gears*

Stereolithography is a process of additive manufacturing, which is used for liquid-based polymers such as epoxy. Berger and Maule (2009) fabricated high reduction epoxy gears by additive manufacturing using stereolithography.

They found the fabrication of harmonic drive and cycloid gears in miniaturized form to be simple and cost-effective. Similarly, Budzik. (2011) successfully fabricated various gear shapes using the stereolithography process.

The fused deposition method (FDM) is a rapid prototyping technique in which a thermoplastic polymer, such as acrylonitrile butadiene styrene, polyamide or polycarbonate, is extruded layer by layer from a nozzle at a controlled temperature. FDM-manufactured polymer gears are widely used in stepper motor planetary gear reducers, linear actuators and positioning devices (Gupta 2018).

In 3DP or inkjet-based systems, a binding agent is injected on the powdered layer to bind the particles to form the shape of a gear (Azari and Nikzad 2009). 3DP has been discussed in more detail by several authors (Torabi *et al.* 2015; Gupta *et al.* (2017). These are different methods used for different polymers to prepare the gear using the additive manufacturing process.

LIGA is an additive manufacturing process that combines lithography, electroplating and molding. It is a German acronym for *Lithographie, Galvanoformung, Abformung*. It is a very useful process for making polymer micro-gears with high aspect ratios. The requirements of an X-ray or ultra-violet light source for the LIGA process make it complicated and expensive. Special protection of the operator and multiple states required to achieve the product are some of the limitations of this process (Jain and Chaubey 2017). However, it provides a good dimensional accuracy and surface finish. For example, micro-pumps and micro-motors are fabricated using the LIGA manufacturing process.

2.4. Laser ablation

This method involves a laser machining operation such as laser beam milling, laser cutting and laser drilling, which is mainly used for manufacturing miniature gears. It uses a laser resource and falls into the substrate that causes the ablation. Ablation uses both vaporization and melt removal when the laser focuses on the substrate. The ablation method is chemical, thermal or a combination of both. Photochemical ablation occurs in the polymer, which breaks the covalent bond of the polymer chain (Ravi-Kumara *et al.* 2019).

The laser ablation process is applied to produce micro-planetary gears and gear tool electrodes for EDM. It is also used for micro-devices for biomedical applications. In this modern era, research on polymer-based laser ablation has been increasing to make polymer gears. However, this manufacturing process causes sharp tooth edges, poor surface finish, thermal distortion and material hardening (Chaubey and Jain 2017).

2.5. Hot embossing

This process is used to manufacture micro-polymer gears only at low cost and slower rates. It is a common pattering technique of microfabrication of polymer gears. The hot embossing process is performed on a hot embossing molding machine. Figure 2.6 shows the basic process of hot embossing for making polymer gears. The basic components of a hot embossing machine can be divided into four major groups (Worgull 2009):

– Press: a frame with an integrated drive unit to provide a molding force and relative movement between the mold insert and the polymer.

– Tool: it is joined together in the molding press. A vacuum unit, heating and cooling unit are an integral part of this tool with an optional alignment system.

– Mold insert: permanently fixed in the hot embossing tool.

– Control system: used to control the process parameter.

Kuo and Zhuang (2014) manufactured a POM gear using this technique. They concluded that process parameters such as hot embossing pressure and embossing temperature influenced the accuracy and performance of the gear. This manufacturing process is used to produce polymer gears with high aspect ratios and low cost.

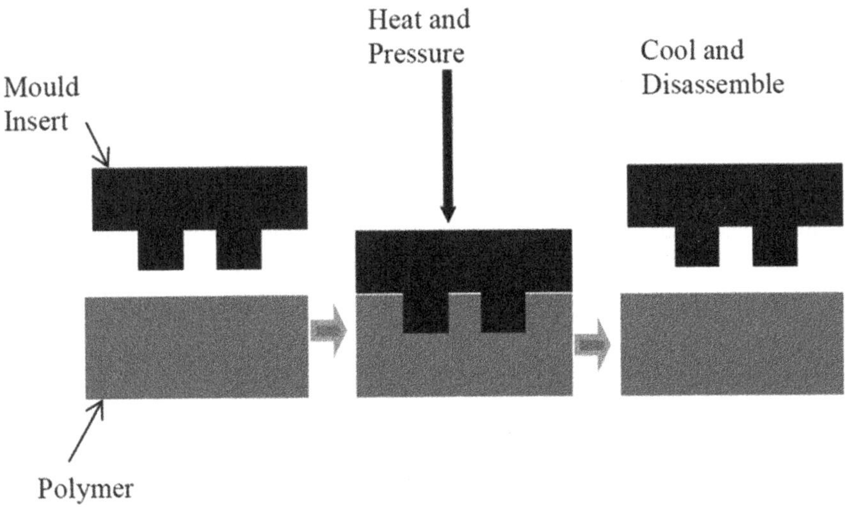

Figure 2.6. *Basic process of hot embossing to manufacture polymer gears*

2.6. Conclusion

Spur gears can be manufactured by different processes such as gear hobbing, injection molding, wire electric discharge machining and the additive manufacturing process. Gear hobbing is a complex manufacturing process in which the injection molding process is a simple technique. Different process parameters such as injection time, packing time, cooling time, packing pressure, melt temperature, mold temperature and injection pressure are important parameters for manufacturing effective polymer gears. Moreover, laser ablation and hot embossing processes are very superior, economical and viable alternatives to manufacturing high-quality miniature gears. However, the additive manufacturing process has the benefits of quality and good surface integrity gears, flexibility, minimum waste, energy efficiency, customer-ordered production and environmental benefits due to low emissions.

2.7. References

Adams, CE. (1986). *Plastic Gearing: Selection and Application*. Marcel Dekker, New York.

Azari, A. and Nikzad, S. (2009). The evolution of rapid prototyping in dentistry: A review. *Rapid Prototyping Journal*, 15, 216–225.

Berger, U. and Maule, B. (2009). Rapid manufacturing of high reduction polymer gears by use of stereolithography. *International Conference on Advanced Intelligent Mechatronics*, Singapore.

Budzik, G. (2011). The use of the rapid prototyping method for the manufacture and examination of gear wheels. In *Advanced Applications of Rapid Prototyping Technology in Modern Engineering*, Hoque, M. (ed.). InTech, London.

Chaubey, S.K. and Jain, N.K. (2017). State-of-art review of past research on manufacturing of meso and micro cylindrical gears. *014 Precision Engineering*. DOI: 10.1016/j.precisioneng.2017.07.

Gupta, K. (2018). Recent developments in additive manufacturing of gears: A review. *16th International Conference on Manufacturing Research ICMR*, Sweden. DOI: 10.3233/978-1-61499-902-7-131.

Gupta, K., Jain, N.K., and Laubscher, R.F. (2017). Advances in gear manufacturing. *Advanced Gear Manufacturing and Finishing*. Academic Press Inc, London.

Jafarzadeh, M. and Farokhi, F. (2016). Design and construction of an automatic syringe injection pump. *Pacific Science Review A: Natural Science and Engineering*, 18(2), 132–137.

Jain, N.K. and Chaubey, S.K. (2017). Review of miniature gear manufacturing. *Reference Module in Materials Science and Materials Engineering Comprehensive Materials Finishing*, 1, 504–538.

Kapelevich, A.L. and McNamara, T.M. (2005). *Direct Gear Design® for Automotive Applications*. SAE International, 05P-149.

Kim, G.H., Lee, J.W., and Seo, T. (2013). Durability characteristics analysis of plastic worm wheel with glass fibre reinforced polyamide. *Materials*, 6, 1873–1890.

Kumar, P., Kommogi, R.K., and Senthilvelan, S. (2009). Injection molded asymmetric spur gear development and preliminary performance evaluation. *International Journal of Plastics Technology*, 13, 186–192.

Kuo, C.-C. and Zhuang, B.-C. (2014). A simple and low-cost method to the development of a micro-featured mold insert for micro-hot embossing. *Indian Journal of Engineering and Materials Sciences*, 21(5), 487–494.

Lee, S.C., Huh, Y.J., Kim, C.H., and Kwon, O.K. (1998). Unified design method of the cavity for injection molded spur gears. *Proceedings of SPE Annual Technical Conference*, Atlanta.

Luscher, A., Houser, D., and Snow, C. (2000). An investigation of the geometry and transmission error of injection molded gear. *Journal of Injection Molding Technology*, 4, 177–190.

Mao, K. (2007). A new approach for polymer composite gear design. *Wear*, 262(3–4), 432–441.

Mao, K., Greenwood, D., Ramakrishnan, R., Goodship, V., Shrouti, C., Chetwynd, D., and Langlois, P. (2019). The wear resistance improvement of fibre reinforced polymer composite gears. *Wear*, 426–427, 1033–1039.

Ravi-Kumara, S., Lies, B., Lyu, H., and Qin, H. (2019). Laser ablation of polymers: A review. *Procedia Manufacturing*, 34, 316–327.

Sardar, J. and Bandopadhya, D. (2014). Development and fabrication of cement reinforced polypropylene composite material spur gear. *Journal of Polymer Engineering*, 34(8), 775–786.

Senthilvelan, S. and Gnanamoorthy, R. (2004a). Damage mechanisms in injection molded unreinforced, glass and carbon reinforced nylon 66 spur gears. *Applied Composite Materials*, 11, 377–397.

Senthilvelan, S. and Gnanamoorthy, R. (2004b). Wear characteristics of injection molded unfilled and glass filled nylon 6 spur gears. *Journal of Engineering Tribology Proceedings of the IMechE – Part J*, 218(6), 495–502.

Senthilvelan, S. and Gnanamoorthy, R. (2006). Fibre reinforcement in injection molded Nylon 6/6 spur gears. *Applied Composite Materials*, 13, 237–248.

Siddhartha, P.K.S. and Singh, A.K. (2018). An investigation on the thermal and wear behavior of polymer based spur gears. *Tribology International*, 118, 264–272.

Stark, S., Beutner, M., Lorenz, F., Uhlmann, S., Karpuschewski, B., and Halle, T. (2013). Heat flux and temperature distribution in gear hobbing operations. *Procedia CIRP*, 8, 456–461.

Sudhagar, S., Mugesh Raja, V., Sathees Kumar, S., and Jonathan Samuel, A. (2019). The wear behavior and service life of Madar and Bauhinia Racemosa reinforced polyester hybrid composites for gear applications. *Materials Today: Proceedings*. Available at: https://doi.org/10.1016/j.matpr.2019.07.738.

Torabi, K., Farjood, E., and Hamedani, S. (2015). Rapid prototyping technologies and their applications in prosthodontics, a review of literature. *Journal of Dental Shiraz University of Medical Sciences*, 16(1), 1–9.

Townsend, D.P. (1991). *Dudley's Gear Handbook*. McGraw-Hill, New York.

Worgull, M. (2009). Hot embossing technique. In *Hot Embossing: Theory and Technology of Microreplication*, Worgull, M. (ed). Elsevier, Burlington, MA, USA.

– introduction of lightweight metal alloys (Mg and Al);

– joining and welding polymer to polymer, metal to metal and metal to polymer;

– industrial application of lightweight metal and polymers.

3.2. Polymer and composites

Chemically bounded long repeating chains of monomers are known as polymers. The chemical reactions involved in the bonding of monomers are known as polymerization. Polymers are categorized as thermoplastics, thermosets, synthetic fibers and elastomers (Vlachopoulos and Strutt 2003). Polymer composites are multiphase solid materials in which one of the phases has one, two or three dimensions. For developing polymer composites, matrix and reinforcement are two essential phases that are generally composed of organic polymers and fibers as the matrix and reinforcement respectively. The strength of reinforcing elements is usually higher than that of the matrix and serves as a load support element. Polymer/plastic is becoming a necessity for humankind. The dependence of humans on plastic is increasing day by day. After a continuous exploration of polymer science, polymers are found to be capable of replacing metals with a very good aspect ratio related to weight-to-strength ratio and good corrosion resistance (Wang *et al.* 2011).

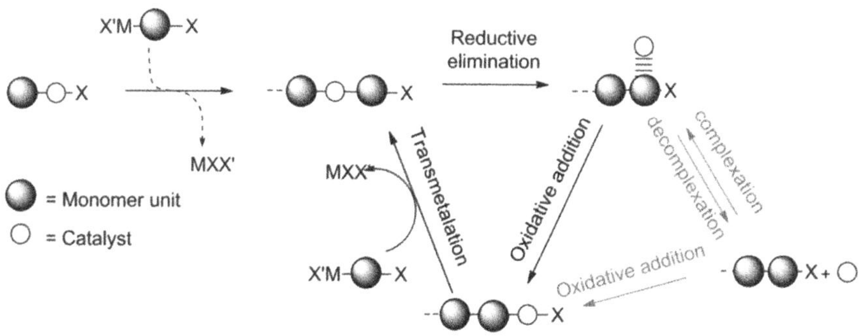

Figure 3.1. *Two ways of controlled polymerization of CPs (Verheyen* et al. *2017)*

3.3. Polymerization

Molecular chemical reactions in monomers to produce a polymer chain or 3D network are known as polymerization. The degree of polymerization and polymer synthesis techniques play a significant role in the performance of polymers and their composites. Polymerization reactions are classified into:

– step-growth reactions; and

– chain growth reactions.

Depending on the method of polymerization, the physical properties of polymers can differ considerably; for example, polymers synthesized by chain-growth polymerization have higher molecular mass (Su 2013).

Step-growth polymerization (SGP): in SGP, only one type of chemical reaction happens, which connects molecules of all sizes. The size of polymer molecules increases at a slow rate over a much longer period of time. The development of a polymer chain follows the processing route; monomer → dimer → trimer → tetramer and so on, until full-sized polymer molecules are produced. During polymerization, polymer chains grow continuously in both ends, and the duration of polymerization is usually in hours. The SGP (Figure 3.2(b)) is categorized as linear, nonlinear and interfacial polymerization.

Interfacial polymerization is also a popular method to produce polymers with high molecular mass. This process requires very fast polymerization. When the polymer molecules start to grow and monomer molecules begin to add to the ends of the polymer chain, the entering monomer molecules tend to react with the ends of polymer chain before they can penetrate through the polymer film to initiate the growth of new chains. As a result, polymers of higher molecular mass are formed. Interfacial polymerization is used to produce flawless, ultrathin films of metal–organic frameworks, bio-hybrids and organic–inorganic hybrids. In nonlinear polymerization, more than two functional groups per molecule are polymerized with the molecules of monomers (Raaijmakers and Benes 2016). Polymer production by nonlinear step-growth polymerization is shown in Figure 3.2(a).

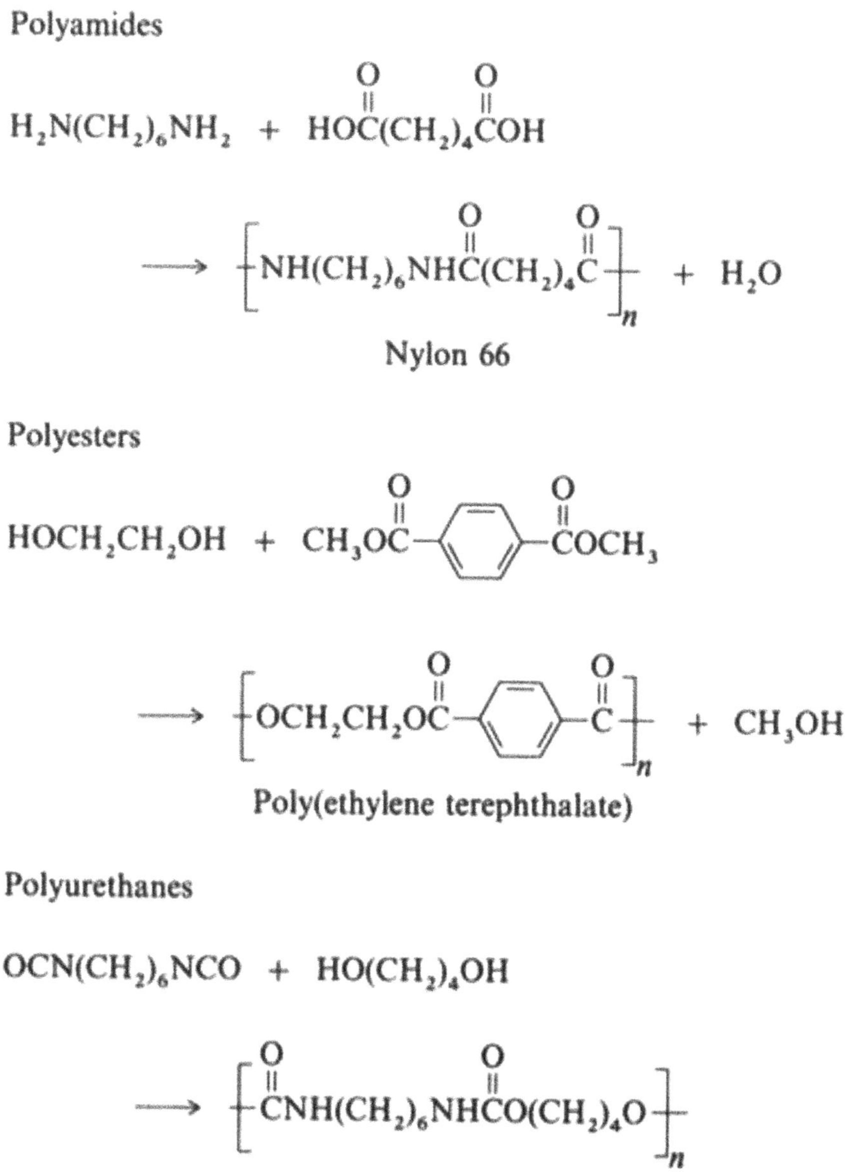

(a)

Figure 3.2a. *Polymers produced by linear step-growth polymerization (Hallensleben 2012)*

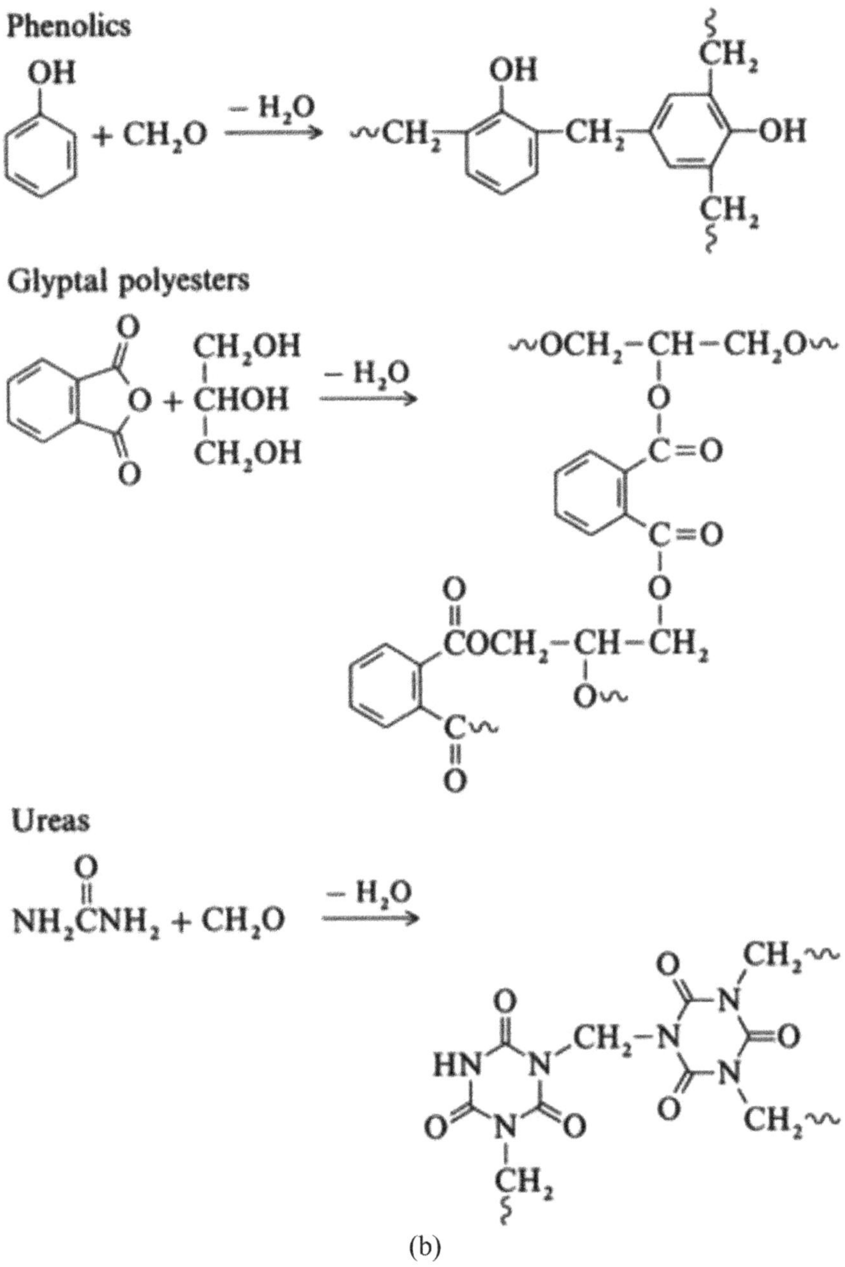

Figure 3.2b. *Polymers produced by nonlinear step-growth polymerization (Hallensleben 2012)*

3.4. Synthesis of polymer composites

The synthesis of polymer composites can be done using various methods, as listed in Table 3.1. The selection of an appropriate method of processing is based on the type of composite to be synthesized, such as a fiber/particulate/nanocomposite. The oldest and easiest method of processing is the hand lay-up technique, and this process is most widely used in basic processing. Components manufactured by injection molding and compression molding having good finishing and strength.

Hand lay-up process (HPL) (Elkington *et al.* 2015)	This is the easiest and most commercialized method of molding also considered as a contact molding process. In this process route of HLP; preparation of mold → apply a thin layer of anti-sticking gel at the base of the mold (for better surface finishes) → apply a reinforcing material like Woven fiber → apply a resin with a curing agent → degas the entire mould. Repeat this processing route multiple times until the desired thickness is achieved. 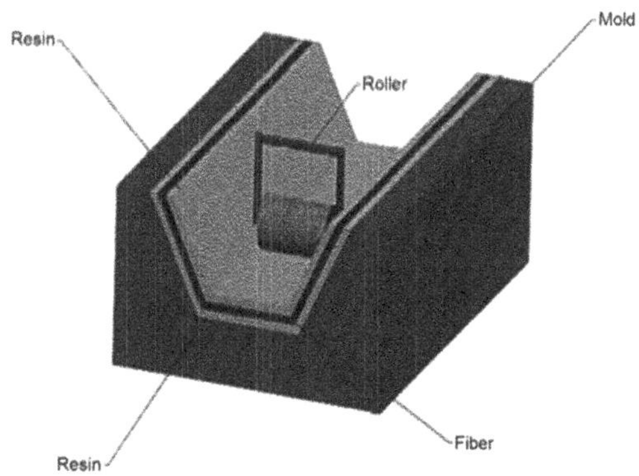
Spray lay-up process (Xiao *et al.* 2015)	Spray lay-up is similar to the hand lay-up process. In this process the matrix and reinforcement are sprayed simultaneously then a roller is applied to remove the air bubbles. The material is sprayed until the desired thickness is achieved. The mould is left for curing at room temperature for several hours. This method is mainly used for manufacturing components of lower load bearing capacity such as boats of truck fairings, small size, bath tubs, etc.

<table>
<tr><td></td><td>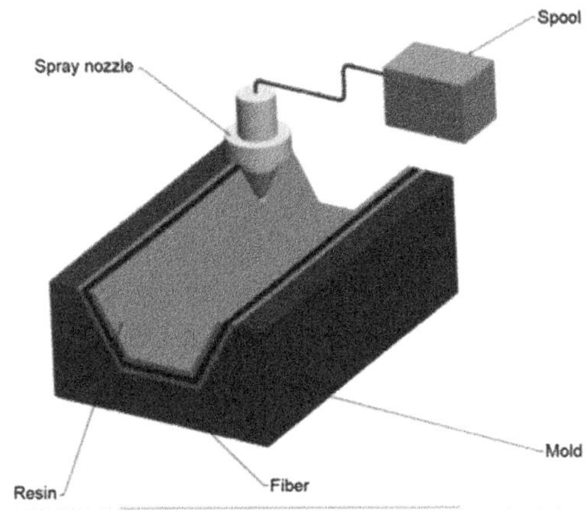
</td></tr>
<tr><td>Compression molding
(Wu et al. 2018)</td><td>In this process the matrix and reinforcement material are kept between the compression die. The appropriate heat and load are applied for the required period of time, the matrix flow in the die with the reinforcement then leave the mold for the required curing time. The strength and quality of the product depend on the type of reinforcement, load, temperature and curing time.
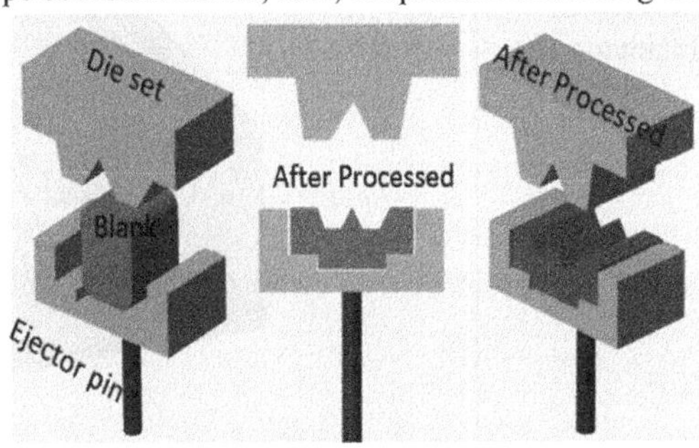
</td></tr>
<tr><td>Resin transfer molding process (Robinson and Kosmatka 2014)</td><td>As shown in the figure the polymeric material is kept in the cylinder and an appropriate heat is applied to melt the polymer then a plunger transfers the liquid material under the influence of compression and gravity. The molten polymer fills the cavity. The product is removed after proper solidification.</td></tr>
</table>

	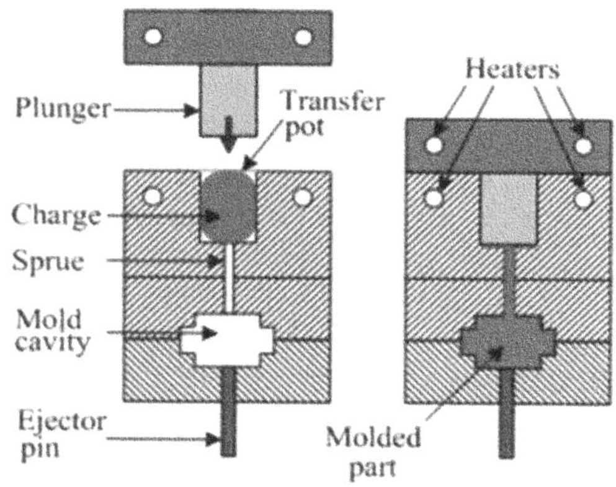
Injection molding process (Singh and Verma 2017)	This technique is usually specially used for thermoplastic polymers to fabricate a product of this surface. As shown in the figure the polymer in granular form is placed in the hopper. The threaded screw with a heating element for melting the polymer, gradually feed the polymer into the cavity. After solidification the product is removed from the mold. 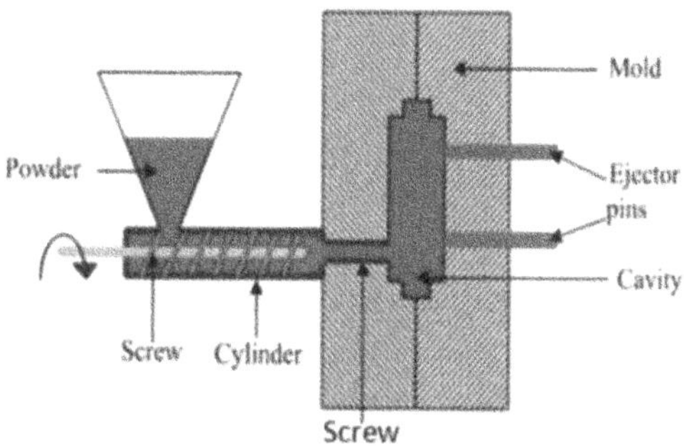
Pultrusion process (Sun *et al.* 2015)	The pultrution is the combination of pull and extrusion. This process is similar to extrusion, the only difference is that in extrusion the material is pushed by piston or plunger but in pultrusion the material is pulled by die.
Filament winding (Quanjin *et al.* 2018)	This technique is suitable for manufacturing hollow cylindrical products like hollow cylindrical vessels, storage tanks for fuel, chemicals and dyes, hollow pipes, rocket motor cases, pressure vessels etc.
Vacuum-bag molding (Bere 2014)	In this process a thin flexible film of nylon, PVA is mounted over a priorly lay-up surface i.e. superimposed with reinforcement. All the sides of the mold are required to be sealed. Atmospheric pressure is used to reduce the air pressure

	from the vacuum bag. The surrounding atmospheric pressure applies a force on the vacuum bag to combine the laminates by removing entrapped air and excess resin, which result in dense fiber composite forms.
Pressure-bag molding (Park *et al.* 2018)	The process is similar to vacuum bag molding; the difference is that the vacuum bag is replaced by pressure plate and the level of air pressure varies from 200 to 300 KPa. This pressure is applied on the rubber sheet to remove excess resin and trapped bubbles and consolidate the reinforcement. This process is commonly used in the manufacture of aircraft fairing, tanks, large containers, wind turbine blades, etc.
Autoclave molding (Centea *et al.* 2015)	This process is capable of manufacturing a void free and dense product because it uses extremely high pressure ranging from 350 to 700 KPa. The process is performed under a completely sealed mold flange and there is no need to apply a thin sheet of polyvinyl alcohol. This technique is more commonly used in aerospace engineering for the fabrication of missile and spacecraft components.

Table 3.1. *Description of various methods for processing polymer composites (Advani and Hsiao 2012; Biswal* et al. *2020)*

3.5. Types of fillers in composites

A filler or reinforcement is added in the polymer (resin: to hold the reinforcement together) to improve its performance such as strength, chemical stability, glass transition temperature and conductivity. If the filler is in the form of fiber/fabrics, the composite is known as fiber composite, and the detailed classification of fibers is shown in Figure 3.3. If the filler is in the form of particles with micro- and nanosize, the composites are considered particulate composites and nanocomposites, respectively. Fibers are generally classified as natural and synthetic fibers (nylon, acrylic, polyethylene, aromatic, aramid polyester, glass, carbon, boron, etc.) (Friedrich *et al.* 2005; Ku *et al.* 2011). The performance of various natural fibers reinforced in epoxy resin is shown in Table 3.2.

Sl. No	Composites	Tensile strength (MPa)	Flexural strength (MPa)	Impact strength (kJ/m^2)
1	Coir-epoxy	23.68	46.63	26.43
2	Sisal-epoxy	37.40	52.8	56.7
3	Jute-epoxy	43.00	55.8	65.00
4	Banana-epoxy	59.00	76.53	149.66
5	Baggase-epoxy	42.40	56.7	110.66
6	Flax-epoxy	59.85	75.40	191.71
7	Areca-epoxy	27.50	25.00	93.33
8	Ramie-epoxy	90.00	110.00	105.40
9	Lantana camara-epoxy	19.08	55.45	32.30
10	Pseudo stem Banana-epoxy	45.57	73.58	92.66
11	Groundnutshell-epoxy	18.09	28.00	24.17
12	Rice husk-epoxy	23.00	29.00	26.00
13	Coir pith-epoxy	9.00	23.00	18.67
14	Banana + sisal-epoxy	25.00	62.00	98.66
15	Luffa + Groundnut- epoxy	39.31	58.95	27.33

Table 3.2. *Natural fiber–epoxy composites – mechanical behavior (Jeyapragash* et al. *2020)*

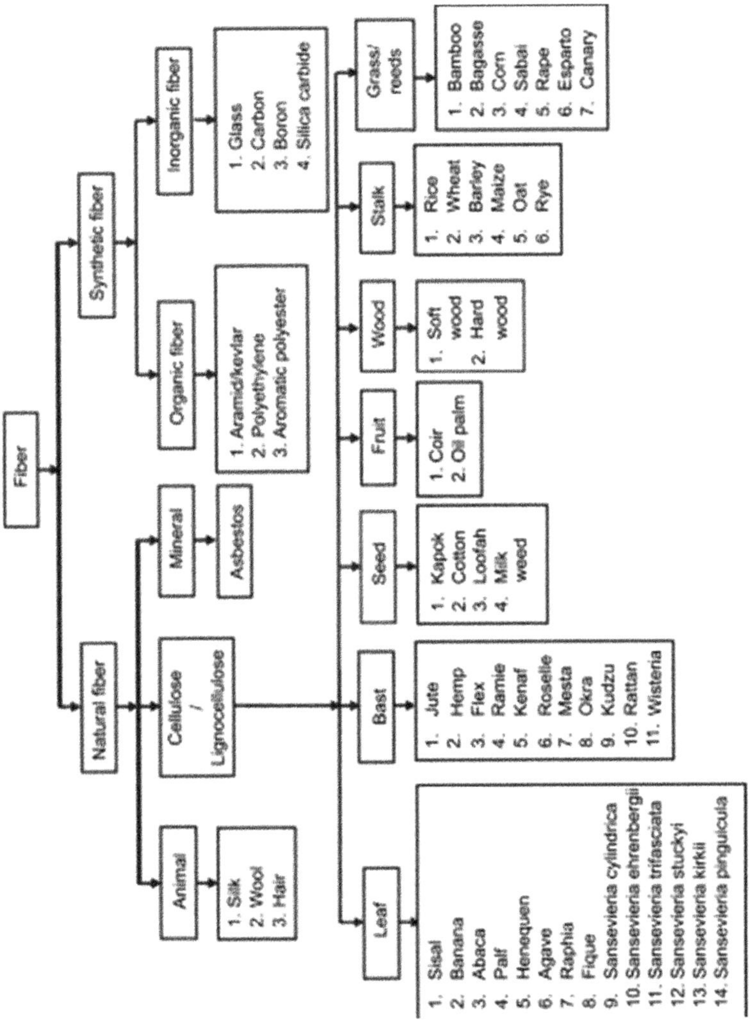

Figure 3.3. *Detailed classification of fiber reinforcement used in polymer composites (Ghalia et al. 2018)*

3.5.1. *Effect of reinforcement orientation*

Particulate composites and chopped or nanofiber composites are randomly distributed without any specified orientation, and the dispersive orientation of this type of reinforcement does not affect the strength, but uniform dispersion of particulates influences the strengthening of composites. One study revealed that poorly dispersed CNTs, compared to well-dispersed CNTs, show higher storage modulus, complex viscosity and loss modulus. Poor dispersion or agglomeration deteriorates the tensile strength and elongation, while tensile modulus remained unaffected (Erden and Ho 2017). Another study reports that the orientation of fiber glass in a continuous manner enhanced the tensile strength from 52 to 92% compared to randomly oriented graphene nanoplatelets, and the impact strength improved by 17–24% using continuous fiber glass compared to random fiberglass (Nassar and Nassar 2020). Similarly, research on the effect of orientation revealed that the tensile strength and flexural strength were 38.84 MPa and 151.22 MPa respectively in 90° orientation (Kumaresan *et al.* 2015).

3.6. Welding polymers

Welding thermoplastic composites is an innovative approach to polymer joining technology. There are various methods of joining polymer composites, and almost all welding methods require heat for fusion. Various joining methods of polymer welding are described in Table 3.3.

Laser and infrared welding	Laser radiation is transmitted through a transparent plate/sheet of polymer, the transmitted radiation is absorbed by a lower plate and heating starts at the interface. The radiant heat at the interface melts the adjacent surface to be welded. The mechanism of heating is due to molecular vibration, and this molecular resonance is responsible for the welding heat. The various areas of LBW application are in medical packaging, electronic and micro-components. The joining of instrument panel clusters and headlights, among others find application in automotive industries.
Microwave welding	Microwave welding is very rapid joining process. In this method a conductive sheet of polymer or metal is placed in between two polymer plates to be joined and placed in the chamber of a microwave. The waves heat the middle plate and melt the interface of the polymer plate applying pressure for fusion. The volumetric heating of waves is the main advantage of the microwave welding process. This process is widely used for welding automobile parts.

Ultrasonic welding	The heat produced for fusion is generated by frictional heating. During this process two plates in lap joint are pressed together superimposed with ultrasonic vibration. After generating adequate heat at the interface a clamping force is applied for fusion and left until solidification occurs. The significant parameters of USW are ultrasonic frequency, amplitude, clamping pressure and cycle time, which affect the weld quality.
Hot plate welding	This is a very conventional method of plastic joining. The parts to be joined are placed in contact with a heated plate, the contact surfaces of the parts become heated to melting. The heated plate is removed and the interfaces of the plates are connected. HPW is used in the pipeline, automotive and construction industries and hydraulic reservoirs.
Hot gas welding	The process is very similar to oxyacetylene gas welding. A gas flame is used to heat and soften the joint area and the filler rod, when the meltingwelding temperature is achieved an adequate intensity of pressure is applied to the joint area to consolidate the joint.
Resistive implant welding	Resistive implant welding uses electric resistance heat. A metallic element or implant is placed at the interface. Heat is produced at the interface by the heating implants, and the faying surface of parts to be joined melts; then with the application of pressure the parts becomes consolidated. The implants or metallic wire become trapped inside the joint. The extensive application of resistive implant welding includes joining of sailing dinghy hulls, car bumpers and other car parts.
Extrusion welding	The process is very similar to injection molding. A filler rod granulate will extrude in the melting or softening stage at the joint groove, and a hot gas flame (for pre heating the parts to be welded) will follow this continuous extrusion until fusion of the filler with the base plate. The molten polymer deposited on the joint line is pressed by a shoe-tool designed in the required bead geometry.
Radio frequency welding	The dielectric hysteretic loss of the polar thermoplastics are key to the association of heating. The dipoles of polar polymers start reoriented due to the influence of a high frequency alternate electric field and produce heat for welding. Hence, the increasing temperature will melt the interface creating a bond, which will consolidate under pressure.
Vibration Welding	The two parts of the polymer to be welded are in contact at the interface. According to the selected parameters the parts start sliding over each other when under pressure. Heat is produced due to friction at the interface. When adequate heat is generated the vibration is terminated and a clamping pressure is applied until the parts become welded.

Table 3.3. *Polymer welding description (Santos 2009)*

3.7. Introduction of lightweight metal and alloys

3.7.1. *Magnesium alloys*

Magnesium is a lightweight and low-density material. It is 33% lighter than aluminum and 77% lighter than steel (Jayasathyakawin *et al.* 2020). Magnesium is most widely used in automobile and aerospace industries due to its lightweight and high-strength characteristics (Xu *et al.* 2017). Magnesium is highly reactive in saline environments, and its reactive nature limits the extension of its applications. It also shows excellent biodegradable properties. The properties of magnesium are very similar to those of bone; it also shows osteoconductivity due to the biodegradable behavior of Mg, and its alloys are most widely used for temporary implants. The Mg alloys used in biological research are categorized as WE (Mg-RE-Zr), AZ (Mg-Al-Zn) and ZK (Mg-Zn-Zr) series alloy (Liu *et al.* 2018).

The low-density and good mechanical characteristics of magnesium provide a wide space to itself in the field of automobile and aerospace industries. The high-performance lightweight Mg alloys are capable of replacing aluminum and steel. Research to enhance the performance of Mg and its alloys is ongoing. Rare earth (RE) elements are emphasized more to improve the strength and corrosion resistance of Mg. One of the RE elements, Gd, is used as an alloying element to improve the heat resistance of cast magnesium alloys (Peng *et al.* 2007). The performance of different alloying elements combined with Gd in cast Mg is shown in Table 3.4.

For the commercialization of RE alloying element Mg alloys is also a great challenge due the high cast of RE elements that limit the commercialization of Mg alloy with RE elements. Therefore, it is necessary to develop a low-cost and high-performance Mg alloy (Table 3.5).

Mg-5Y shows a good corrosion resistance with relatively less cost compared to pure Mg (Cao *et al.* 2018). The addition of 10 (wt%) Zn and 5 (wt%) Al in Mg developed a high tensile and fatigue property (Mohammed *et al.* 2019).

Alloy (wt%)	Cast method	Heat treatment	YS (MPa)	UTS (MPa)	EL (%)
Mg-10Gd-1Zn-0.5Zr	Permanent mold cast	460 °C × 12 h + 200 °C × 48h	205	303	6.6
Mg-10Gd-2Y-1Zn-0.5Zr	Permanent mold cast	480 °C × 12 h + 200 °C × 12h	252	351	10.2
Mg–12Gd–0.8Zn–0.4Zr	Sand cast	530 °C × 18 h + 225 °C × 8 h	–	348	2.6
Mg-10Gd-3Y-0.5Zr	Sand cast	525 °C × 12 h + 250 °C × 12 h	–	325	5
Mg–2Zn–0.2Y–0.5Nd–0.4Zr	Permanent mold cast	As-cast	89	203	35%

Table 3.4. *Mechanical performance of cast Mg alloys at room temperature (Song* et al. *2020)*

Alloy (wt%)	Cast method	YS (MPa)	UTS (MPa)	EL (%)	Remark
Mg-5Y	Permanent mold cast	–	–	–	Corrosion resistant
Mg-10Zn-5Al	High pressure die cast	188	235	5.6	Superior fatigue property
Mg–3.5Al–4.2La–0.3Mn	High pressure die cast	–	–	–	Better property than A380
Mg–1Nd–1Ce-0.5Zn-Zr	Permanent mold cast	136	237	9	–

Table 3.5. *Properties of low-cost, high-performance cast Mg alloys at room temperature (Song* et al. *2020)*

It was reported that alloying of Ca, Ce and Sr does not affect the mechanism of deformation under the influence of temperature and strain rate. Alloying Ca leads to perceptible grain refinement and a more homogeneous microstructure and results in an increase in the ductility of the AZ31 alloy at high temperature and low strain rate conditions. It was reported that the elongation to failure of AZ31 is 347%, however when alloyed with Ca, it improved to 406%; alloying Ca and Ce improved it to 437% and alloying Ca, Sr and Ce improved it to 552% (Shang *et al.* 2011; Kwak *et al.* 2016). Alloying Li is also a great potential element, which is capable of optimizing the microstructure of the Mg–Al alloy (You *et al.* 2017).

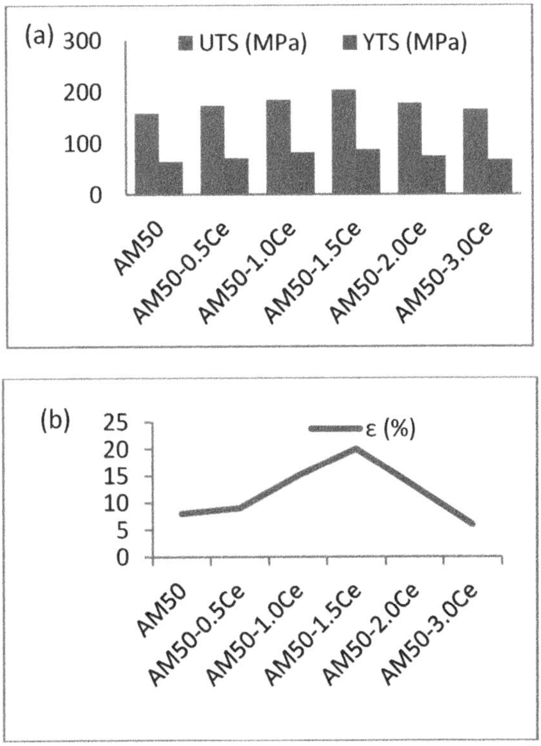

Figure 3.4. *(a) Effect of percentage addition of Ce on UTS & YTS. (b) Effect of percentage addition of Ce on elongation (You* et al. *2017). For a color version of this figure, see www.iste.co.uk/kumar/materials.zip*

The AM series Mg alloys with basic alloying elements Al and Mn are AM50 and AM60. The AM series alloy offers good corrosion resistance, thus finding applications in instrument panels and steering wheels. The major concern for AM series alloys are their poor performance at high temperatures. Therefore, these alloys can be used only at 120°C (Wang *et al.* 2009). Addition of alloying elements, such as Ce, can improve the performance of Am series alloys, as shown in Figures 3.4(a) and (b).

3.7.2. Aluminum alloys

Aluminum is the most abundant and third most common metal in the world. It is the second most widely used metal in the world after steel. The excellent properties of Al make it a versatile, economical and attractive metal. Al has perfect features for various fields, such as high electrical conductivity, high ductility, high strength, high thermal conductivity, good formability, excellent corrosion resistance, ease of fabrication, bright and attractive color and texture, and most importantly, it is a lightweight metal with a high strength-to-density ratio. With all of the above features, Al alloys are best suited to structural and industrial applications such as shipbuilding and marine industries, automotive, aerospace and rail industries, due to their tendency to reduce weight and enhance efficiency (Kammer 2018). Al is the most commonly used material in aircraft industries, and constitutes about 66% of the total empty weight of a recent passenger aircraft manufactured by airbus (Starke and Staley 2010; Al Jawalkar and Kant 2015). The percentages of aluminum used in the construction of aircraft manufactured by various industries are shown in Table 3.8.

Aluminum alloys are classified according to the Aluminum Association standards. The wrought Al alloys have a four digit designation system (XXXX) in which one series are non-heat treatable, i.e. 1XXX, 3XXX and 5XXX, and can be processed by cold working. Another series, the 2XXX, 6XXX, 7XXX and 8XXX group of alloys, are age hard-enabled (Azarniya *et al.* 2019). The detailed description of Al alloys with their main alloying elements, properties and application is shown in Table 3.7.

Al alloys	Main alloying element	Properties	Application
1 XXX	Pure Al (99% or greater)	1 Corrosion resistive 2 Excellent ductility 3 Low mechanical strength	Most commonly used for power transmission lines that connect the national grids, 1350 for electrical applications and 1100 for food package trays.

2 XXX	Al–Cu alloys	1 High strength 2 Higher weldability 3 Excellent fracture toughness	Alloy 2024 is the most widely known aircraft alloy, Al sheets.
3 XXX	Al–Mn alloys	1 Excellent ductility 2 Formability 3 Low strength 3 Excellent corrosion resistance	Heat exchangers, cooking utensils, manufacturing of aluminum beverage cans (alloy 3004) and automotive applications.
4 XXX	Al–Si alloys	1 Admirable formability and ductility 2 Moderate mechanical strength 3 Poor thermal stability due to low fusion point	For marine and building applications in the form of sheets, wire manufacturing, alloy 4043 is predominantly used as a filler alloy for 6xxx series alloy welding.
5 XXX	Al–Mg alloys	1 Appropriate weldability 2 Medium strength 3 Resistance to corrosion	Anodized sheets of Al5005 for buildings; Al5083 is widely used for marine, shipbuilding, aerospace, transport applications; 5052 in electronics; Fighting vehicle is made with 5083 and 7XXX series.
6 XXX	Al–Mg–Si alloys	1 Ductile in nature 2 Moderate strength 3 Formability is good 4 Excellent resistance to corrosion	Used in truck and marine frames; 6xxx series as extruded alloy is the first preference for architectural application; iPhone 6 extrusion made by 6th series alloy.
7 XXX	Al–Zn–Mg–Cu alloys	1 High strength 2 Extensive vulnerability to stress corrosion	7050 and 7075 are widely used by aircraft industries.
8 XXX	Al + other elements	3 Moderate strength	Widely used for bearings manufacturing.
9 XXX	Unused series	--	--

Table 3.7. *Designation and description of wrought and cast Al alloys according to the Aluminum Association standard designation system (The Aluminum Association Inc. 2006; Azarniya* et al. *2019)*

<table>
<tr><th>Aircraft</th><th>Aluminum</th><th>Steel</th><th>Titanium</th><th>PMCs</th><th>Other</th></tr>
<tr><td>Boeing 747</td><td>81 wt.%</td><td>13 wt.%</td><td>4 wt.%</td><td>1 wt.%</td><td>1 wt.%</td></tr>
<tr><td>Boeing 757</td><td>78 wt.%</td><td>12 wt.%</td><td>6 wt.%</td><td>3 wt.%</td><td>1 wt.%</td></tr>
<tr><td>Boeing 767</td><td>80 wt.%</td><td>14 wt.%</td><td>2 wt.%</td><td>3 wt.%</td><td>1 wt.%</td></tr>
<tr><td>Boeing 777</td><td>70 wt.%</td><td>11 wt.%</td><td>7 wt.%</td><td>11 wt.%</td><td>1 wt.%</td></tr>
<tr><td>Boeing 787</td><td>20 wt.%</td><td>10 wt.%</td><td>15 wt.%</td><td>50 wt.%</td><td>5 wt.%</td></tr>
<tr><td>DC-10</td><td>78 wt.%</td><td>14 wt.%</td><td>5 wt.%</td><td>1 wt.%</td><td>2 wt.%</td></tr>
<tr><td>MD-11</td><td>76 wt.%</td><td>9 wt.%</td><td>5 wt.%</td><td>8 wt.%</td><td>2 wt.%</td></tr>
<tr><td>MD-12</td><td>70 wt.%</td><td>8 wt.%</td><td>4 wt.%</td><td>16 wt.%</td><td>2 wt.%</td></tr>
<tr><td>Airbus A380</td><td>61 wt.%</td><td colspan="2">10 wt.%</td><td>22 wt.%</td><td>7 wt.%</td></tr>
<tr><td>Airbus A350</td><td>20 wt.%</td><td>7 wt.%</td><td>14 wt.%</td><td>52 wt.%</td><td>7 wt.%</td></tr>
</table>

Table 3.8. *Structural material usage in large commercial aircraft (Starke and Staley 2010)*

3.8. Welding dissimilar metal alloys

Joining of dissimilar metals by fusion welding processes such as TIG, laser welding, gas welding and arc welding is highly challenging due to different physical properties such as melting point and conductivity. On the other hand, solid-state welding processes such as friction stir welding and linear friction welding reported the capability to produce a sound joint for welding Al to Cu, Al to Mg and Al to Ti. The recent research continues to investigate the performance of hybrid methods. The most popular techniques for welding dissimilar metals are friction stir welding, magnetic pulse welding and laser (dual) welding. FSW has wide applications in marine, aerospace, automobile and high-speed train and ship industries (Kumar *et al.* 2015).

3.8.1. *Friction stir welding*

Many researchers have reported the capability of the FSW process to weld dissimilar metals. FSW also supports green technology because this process is free from fumes/gases from burning shielding gas and the fusion of flux and metals. The strength of a weld joint depends on the formation of intermetallic compound (IMC) and mechanical interlocking. The process description of FSW is shown in Figure 3.5, and the main process parameters of FSW are rotational speed, geometry of the pin (diameter, length and profile of the pin) and the base of previous research that traverse speed does not significantly affect the quality of the weld joint (Singh *et al.* 2020). During the welding of Al and Mg by FSW, the IMC formation of $A_{13}Mg_2$ and $Al_{12}Mg_{17}$ in the Al + Mg binary phase system is due to the eutectics of Al + Al_3Mg_2 (37% Mg) at 437°C and Mg + $Al_{12}Mg_{17}$ (57% Mg) at 450°C respectively. The strength of the joint also depends on the size and dispersion of IMC. The larger thickness of IMC reduces the joint strength, and the ultrasonic-assisted FSW decreased the bilayer IMC into a monolayer IMC with a decreased overall thickness (Lv *et al.* 2017). The probe/tool offset from the weld center line also affects the joint characteristics. In another study, a 1.5 mm offset probe improves the tensile strength to 69% from the base metal ZK60 alloy, i.e. higher than that of a Ti and Mg joint (Aonuma and Nakata 2012).

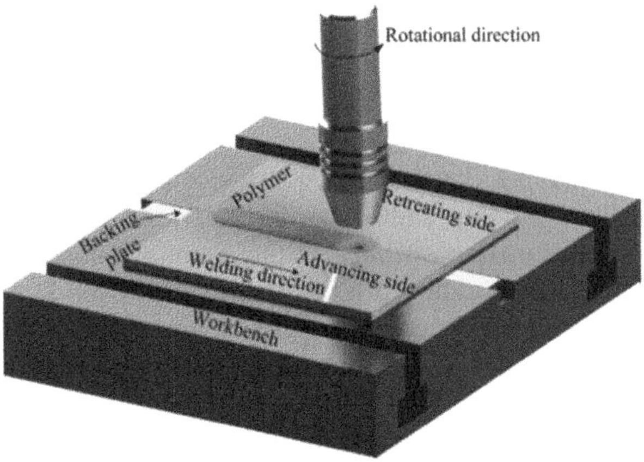

Figure 3.5. *Process description of FSW (Huang* et al. *2018b). For a color version of this figure, see www.iste.co.uk/kumar/materials.zip*

3.8.2. *Welding polymer and metal alloys*

Joining metals and polymers is a critical and trending issue in the current scenario of lightweight application. Metals and polymers have a great difference in thermo-mechanical and thermo-chemical properties. Therefore, homogeneous fusion at the interface using a fusion welding processes is near impossible. The new innovative approach of interlocking by FSW and ultrasonic welding (USW) is continuously accelerating. Many researchers are trying to explore the possibility of joining polymers to metal to fill the gap. The formation of a heterojunction between polymer matrix composites and metals by friction self-riveting welding (FSRW) was performed to join the glass fiber polyphenylene sulfide (GFR-PPS) AA2060-T8. In this process, a pinless tool was used. The metal plate with a hole is placed over the polymer sheet. A high-speed rotating tool travels over the metal plate and heats the metal surface, and the metal transmits the heat to the polymer, and by applying a plunging pressure, the polymer squeezes into the hole in the plate and interlocks (Huang *et al.* 2018a; Meng *et al.* 2019b). The USW is capable of joining the metal and polymer composites, but the process is limited by less than 1 mm sheet thickness. The PE bonded with Mg alloy on the molecular level by using plasma electrolyte oxidation interface produces a strong joint between CFRP and AA5052 by Mg oxide interlayer (Nagatsuka *et al.* 2015). The high-speed friction welding method is investigated to join the Al-alloy 6061-T6 and polyamide 66 (PA66) sheets and high bond strength is found at very high welding speeds (Liu *et al.* 2020). The joining of FRP composites is also challenging because drilling or melting the fiber in the matrix can result in damage. There are various methods of joining FRP with metals, and a few of them are discussed below.

3.8.2.1. *Conventional resistance welding*

In this process, a metal plate with carrier plate is assembled with the FRP plate by application of a force, according to the second step in Figure 3.6; then, after mounting another plate of metal over the FRM plate, the entire assembly is placed between two electrodes for resistance welding. After a supply of power the pin gets heated by electrical resistance and deforms to make multiple interlocks (Reisgen *et al.* 2020).

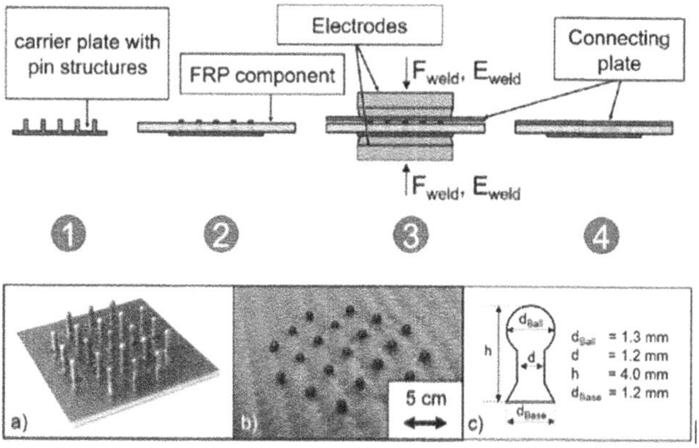

Figure 3.6. *Process route of conventional resistance welding of metal and FRP composite (Reisgen* et al. *2020). For a color version of this figure, see www.iste.co.uk/kumar/materials.zip*

3.8.2.2. *Self-piercing rivet (SPR) technique (Kang* et al. *2016)*

This is a type of deformation process from punching in the die. The die has a space to allow the deformation of parts to be joined. A rivet mounted with punch is pressed and inserted into the plate and deforms in a specific geometry in the die to form an interlock. This process does not require a pre-drilled hole in the plate or proper alignment of the rivet. This is a very fast process because no preparation of the workpiece is required; the processing of the SPR method is shown in Figure 3.7.

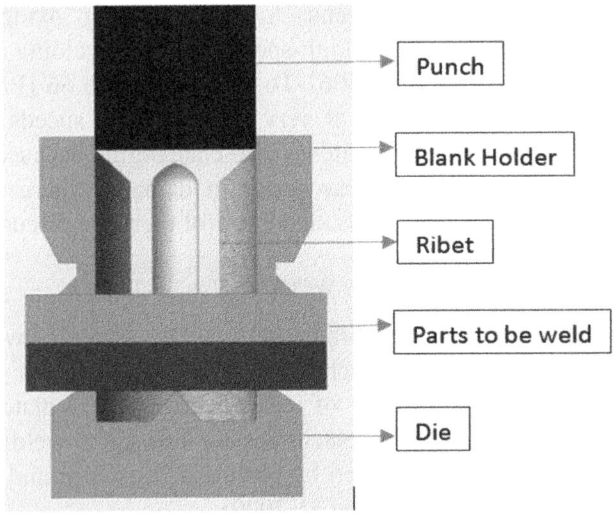

Figure 3.7. *Self-piercing rivet (SPR) technique (Kang* et al. *2016). For a color version of this figure, see www.iste.co.uk/kumar/materials.zip*

3.9. Industrial application of polymers

The participation of plastic/polymers in cars is approximately 350 kg in 2020, which has increased from 200 kg in 2014. It was also reported that CFPC and PMC are capable of reducing the car weight by 25–70%. Recent studies continue to replace all the glass in cars with polycarbonate (PC) [https://www.plasticstoday.com/automotive-and-mobility/plastics-use-vehicles-grow-75-2020-says-industry-watcher/63791493722019]. Different types of polymers and their possible application are shown in Figure 3.8. The other materials used in automobile manufacturing, such as steel, aluminum, magnesium and polymer composites, are discussed with their advantages, applications and drawbacks in Table 3.9.

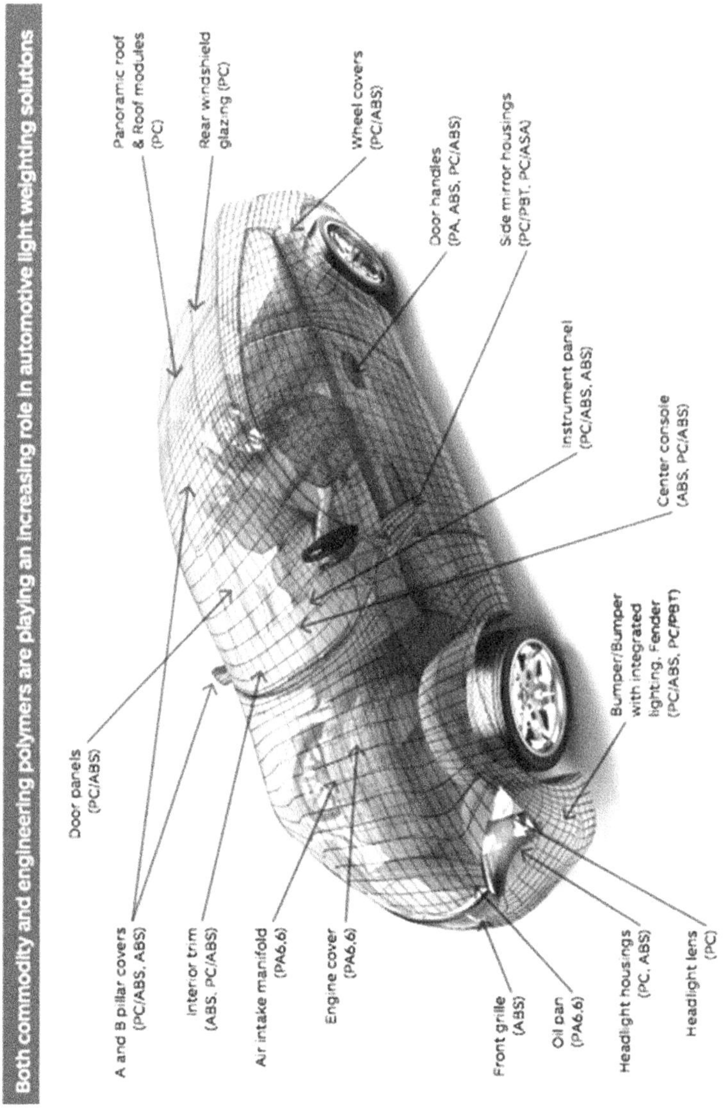

Figure 3.8. *Plastic/polymer composites used in cars (https://blog.americanchemistry.com/wp-content/uploads/2014/11/Automobile-Light-weighting-solutions.png). For a color version of this figure, see www.iste.co.uk/kumar/materials.zip*

Materials	Areas of application	Advantages	Current drawbacks
Steel	Chassis, BIW, power trains, bumpers and engine parts	Strong, high toughness, corrosion resistive (after treatment), good formability, weldability	Very heavy compared to other materials used in automobiles
Aluminum	Engine block, body panels, power trains, piston and cylinder head	Lightweight, corrosion resistive, energy efficient, good formability	Poor performance at high temperature
Magnesium	Inner door structure, steering wheel, steering column, car sheet frame, etc.	Lightest to all other engineering metals, fuel economy efficient	Low melting point, highly reactive, poor corrosion resistance and wear resistance
Copper	Electro-mechanical application, ABS braking system	Good electrical conductivity, compatible for electrical accessories in automobile	Limited application in automobile parts
Composites and reinforced plastics	Passenger cell, roof, wheel rims, hood, etc.	High strength-to-weight ratio and good stiffness, corrosion resistive	Expensive, complex manufacturing process, not recyclable

Table 3.9. *Comparison of different materials used in automobiles [http://article.sapub.org/image/10.5923.j.ijme.20180803.02_001.gif]*

3.10. Conclusion

Polymer composites have inherent properties to replace heavy and low-efficiency metals such as steel. The processing technique and mechanism of polymerization significantly affects the structure and strength of polymer composites. As discussed, step-growth and chain-growth polymerization are the criteria of selection according to the type of monomers and initiators. This chapter discussed the various methods of polymer processing and welding and provided glimpses and understanding of how to select the appropriate process for the type of polymer and reinforcement. Lightweight metals and composites such as magnesium, aluminum and polymer composites have been discussed with regard to various alloying elements to enhance performance and hidden opportunities to improve the desired properties of the rare earth elements that significantly affect performance. Welding a polymer to a metal still challenges researchers who continue to explore innovative techniques to create a sound joint.

3.11. References

Advani, S.G. and Hsiao, K.T. (2012). *Manufacturing Techniques for Polymer Matrix Composites (PMCs)*. Woodhead Publishing, Cambridge.

Aonuma, M. and Nakata, K. (2012). Dissimilar metal joining of ZK60 magnesium alloy and titanium by friction stir welding. *Materials Science and Engineering B*, 177(7), 543–548.

Azarniya, A., Taheri, A.K., and Taheri, K.K. (2019). Recent advances in ageing of 7xxx series aluminum alloys: A physical metallurgy perspective. *Journal of Alloys and Compounds*, 781, 945–983.

Bere, P. (2014). Experimental research regarding vacuum bag technology for obtaining carbon/epoxi composites. *Academic Journal of Manufacturing Engineering*, 12(1), 86–90.

Bhin Park, S., Woo Lee, D., and Il Song, J. (2018). Fabrication and evaluation of mechanical properties of carbon/epoxy square tube using pressure bag molding and compared with autoclave method. *International Journal of Precision Engineering and Manufacturing*, 19(3), 441–446.

Biswal, T., BadJena, S.K., and Pradhan, D. (2020). Synthesis of polymer composite materials and their biomedical applications. *Materials Today: Proceedings*, 1–11, ISSN 2214-7853. Available at: https://doi.org/10.1016/j.matpr.2020.01.567.

Cao, F., Zheng, D., Song, G.L., Shi, Z., and Atrens, A. (2018). The corrosion behavior of Mg5Y in nominally distilled water. *Advanced Engineering Materials*, 20(6), 1700986.

Centea, T., Grunenfelder, L.K., and Nutt, S.R. (2015). A review of out-of-autoclave prepregs – Material properties, process phenomena, and manufacturing considerations. *Composites Part A: Applied Science and Manufacturing*, 70, 132–154.

Elkington, M., Bloom, D., Ward, C., Chatzimichali, A., and Potter, K. (2015). Hand layup: Understanding the manual process. *Advanced Manufacturing: Polymer & Composites Science*, 1(3), 138–151.

Erden, S. and Ho, K. (2017). Fiber reinforced composites. In *Fiber Technology for Fiber-Reinforced Composites*, Seydibeyoğlu, M.O., Mohanty, A.K., and Misra M. (eds). Woodhead Publishing, Cambridge.

Friedrich, K., Zhang, Z., and Schlarb, A.K. (2005). Effects of various fillers on the sliding wear of polymer composites. *Composites Science and Technology*, 65(15–16), 2329–2343.

Gao, L. and Luo, A.A. (2013). Hot deformation behavior of as-cast Mg–Zn–Mn–Ce alloy in compression. *Materials Science and Engineering A*, 560, 492–499.

Ghalia, M.A. and Abdelrasoul, A. (2018). Compressive and fracture toughness of natural and synthetic fiber-reinforced polymer. In *Mechanical and Physical Testing of Biocomposites, Fibre-Reinforced Composites and Hybrid Composites*, Jawaid, M., Thariq, M., and Saba, N, (eds). Woodhead Publishing, Cambridge.

Hallensleben, M.L. (2012). Polyester resins, unsaturated. *Ullmann's Encyclopedia of Industrial Chemistry*, 29, 6.5–621.

Huang, Y., Meng, X., Wang, Y., Xie, Y., and Zhou, L. (2018a). Joining of aluminum alloy and polymer via friction stir lap welding. *Journal of Materials Processing Technology*, 257, 148–154.

Huang, Y., Meng, X., Xie, Y., Wan, L., Lv, Z., Cao, J., and Feng, J. (2018b). Friction stir welding/processing of polymers and polymer matrix composites. *Composites Part A: Applied Science and Manufacturing*, 105, 235–257.

Jawalkar, C.S. and Kant, S. (2015). A review on use of aluminium alloys in aircraft components. *i-manager's Journal on Material Science*, 3(3), 33–38.

Jayasathyakawin, S., Ravichandran, M., Baskar, N., Anand Chairman, C., and Balasundaram, R. (2020). Mechanical properties and applications of magnesium alloy – Review. *Materials Today: Proceedings*, 27.

Jeyapragash, R., Srinivasan, V., and Sathiyamurthy, S. (2020). Mechanical properties of natural fiber/particulate reinforced epoxy composites – A review of the literature. *Materials Today: Proceedings*, 22(3), 1223–1227.

Kammer, C. (2018). *Aluminum and Aluminum Alloys*. Springer Handbooks, New York.

Kang, J., Rao, H., Zhang, R., Avery, K., and Su, X. (2016). Tensile and fatigue behaviour of self-piercing rivets of CFRP to aluminium for automotive application. *IOP Conference Series: Materials Science and Engineering*, 137.

Kawahito, Y. and Katayama, S. (2010). Innovation of laser direct joining between metal and plastic. *Transactions of the JWRI*, 39, 50–52.

Ku, H., Wang, H., Pattarachaiyakoop, N., and Trada, M. (2011). A review on the tensile properties of natural fiber reinforced polymer composites. *Composites Part B-Engineering*, 42, 856–873.

Kumar, N., Yuan, W., and Mishra, R.S.R.S. (2015). *Friction Stir Welding of Dissimilar Alloys and Materials*. Butterworth-Heinemann, Oxford.

Kumaresan, M., Sathish, S., and Karthi, N. (2015). Effect of fiber orientation on mechanical properties of sisal fiber reinforced epoxy composites. *Journal of Applied Science and Engineering*, 18(3), 289–294.

Kwak, T.Y., Lim, H.K., and Kim, W.J. (2016). The effect of 0.5 wt.% Ca addition on the hot compressive characteristics and processing maps of the cast and extruded Mg–3Al–1Zn alloys. *Journal of Alloys and Compounds*, 658, 157–169.

Liu, C., Ren, Z., Xu, Y., Pang, S., Zhao, X., and Zhao, Y. (2018). Biodegradable magnesium alloys developed as bone repair materials: A review. *Scanning*, 2018.

Liu, F.C., Dong, P., and Pei, X. (2020). A high-speed metal-to-polymer direct joining technique and underlying bonding mechanisms. *Journal of Materials Processing Technology*, 280, 116610.

Lv, X.Q., Wu, C.S., and Padhy, G.K. (2017). Diminishing intermetallic compound layer in ultrasonic vibration enhanced friction stir welding of aluminum alloy to magnesium alloy. *Materials Letters*, 203, 81.

Meng, F., Lv, S., Qiang, Y., Qin, P., Zhang, J., Guan, K., Huang, Y., Hort, N., Li, B., Liu, X., and Meng, J. (2019a). Developing a die casting magnesium alloy with excellent mechanical performance by controlling intermetallic phase. *Journal of Alloys and Compounds*, 795, 436–445.

Meng, X., Huang, Y., Xie, Y., Li, J., Guan, M., Wan, L., Dong, Z., and Cao, J. (2019b). Friction self-riveting welding between polymer matrix composites and metals. *Composites Part A: Applied Science and Manufacturing*, 127, 105624.

Mohammed, S.M.A.K., Li, D.J., Zeng, X.Q., and Chen, D.L. (2019). Cyclic deformation behavior of a high zinc-containing cast magnesium alloy. *International Journal of Fatigue*, 125, 1–10.

Nagatsuka, K., Yoshida, S., Tsuchiya, A., and Nakata, K. (2015). Direct joining of carbon-fiber-reinforced plastic to an aluminum alloy using friction lap joining. *Composites Part B: Engineering*, 73, 82–88.

Nassar, A. and Nassar, E. (2020). Effect of fiber orientation on the mechanical properties of multi layers laminate nanocomposites. *Heliyon*, 6(1), e03167.

Peng, Q.M., Wu, Y.M., Fang, D.Q., Meng, J., and Wang, L.M. (2007). Microstructures and properties of Mg-7Gd alloy containing Y. *Journal of Alloys and Compounds*, 430(1–2), 252–256.

Quanjin, M., Rejab, M.R.M., Kaige, J., Idris, M.S., and Harith, M.N. (2018). Filament winding technique, experiment and simulation analysis on tubular structure. *IOP Conference Series: Materials Science and Engineering*, 342.

Raaijmakers, M.J.T. and Benes, N.E. (2016). Current trends in interfacial polymerization chemistry. *Progress in Polymer Science*, 63, 86–142.

Reisgen, U., Schiebahn, A., Lotte, J., Hopmann, C., Schneider, D., and Neuhaus, J. (2020). Innovative joining technology for the production of hybrid components from FRP and metals. *Journal of Materials Processing Technology*, 282, 116674.

Robinson, M.J. and Kosmatka, J.B. (2014). Analysis of the post-filling phase of the vacuum-assisted resin transfer molding process. *Journal of Composite Materials*, 48(13), 1547–1559.

Santos, J.F. (2009). Joining of polymers and polymer – Metal hybrid structures: Recent developments and trends. *Polymer Engineering and Science*, 49(8), 1461–1476.

Shang, L., Yue, S., Verma, R., Krajewski, P., Galvani, C., and Essadiqi, E. (2011). Effect of microalloying (Ca, Sr, and Ce) on elevated temperature tensile behavior of AZ31 magnesium sheet alloy. *Materials Science and Engineering: A*, 528(10–11), 3761–3770.

Singh, G. and Verma, A. (2017). A brief review on injection moulding manufacturing process. *Materials Today: Proceedings*, 4(2), 1423–1433.

Singh, V.P., Patel, S.K., Ranjan, A., and Kuriachen, B. (2020). Recent research progress in solid state friction-stir welding of aluminium-magnesium alloys: A critical review. *Journal of Materials Research and Technology*, 9(3), 1–40.

Song, J., She, J., Chen, D., and Pan, F. (2020). Latest research advances on magnesium and magnesium alloys worldwide. *Journal of Magnesium and Alloys*, 8(1), 1–41.

Starke, E.A. and Staley, J.T. (2010). *Application of Modern Aluminium Alloys to Aircraft*. Woodhead Publishing, Cambridge.

Su, W.-F. (2013). *Principles of Polymer Design and Synthesis. Extracts*. Springer-Verlag Berlin Heidelberg, Berlin.

Sun, S.P., Wei, M., Olson, J.R., and Shaw, M.T. (2015). A modified pultrusion process for preparing composites reinforced with continuous fibers and aligned hydroxyapatite nano needles. *Polymer Composites*. Available at: https://doi.org/10.1002/pc.23013.

The Aluminum Association Inc. (2006). *International Alloy Designations and Chemical Composition Limits for Wrought Aluminum and Wrought Aluminum Alloys*. The Aluminum Association, Arlington, Virginia.

Verheyen, L., Leysen, P., Van Den Eede, M.P., Ceunen, W., Hardeman, T., and Koeckelberghs, G. (2017). Advances in the controlled polymerization of conjugated polymers. *Polymer (Guildf)*, 108, 521–546.

Vlachopoulos, J. and Strutt, D. (2003). Polymer processing. *Materials Science and Technology*, 19, 1161–1169.

Wang, J., Liao, R., Wang, L., Wu, Y., Cao, Z., and Wang, L. (2009). Investigations of the properties of Mg–5Al–0.3Mn-xCe (x = 0–3, wt.%) alloys. *Journal of Alloys and Compounds*, 477(1–2), 341–345.

Wang, R.M., Zheng, S.R., and Zheng, Y.P. (2011). *Polymer Matrix Composites and Technology*. Woodhead Publishing, Cambridge.

Wu, M.S., Centea, T., and Nutt, S.R. (2018). Compression molding of reused in-process waste–effects of material and process factors. *Advanced Manufacturing: Polymer & Composites Science*, 4(1), 1–12.

Xiao, B., Yang, Y., Wu, X., Liao, M., Nishida, R., and Hamada, H. (2015). Hybrid laminated composites molded by spray lay-up process. *Fibers and Polymers*, 16, 1759–1765.

Xu, C., Nakata, T., Qiao, X.G., Jiang, H.S., Sun, W.T., Chi, Y.C., Zheng, M.Y., and Kamado, S. (2017). Effect of extrusion parameters on microstructure and mechanical properties of Mg-7.5Gd-2.5Y-3.5Zn-0.9Ca-0.4Zr (wt%) alloy. *Materials Science and Engineering: A*, 685, 159–167.

You, S., Huang, Y., Kainer, K.U., and Hort, N. (2017). Recent research and developments on wrought magnesium alloys. *Journal of Magnesium and Alloys*, 5(3), 239–253.

Zhou, Y., Fu, P., Peng, L., Wang, D., Wang, Y., Hu, B., Liu, M., Sachdev, A., and Ding, W. (2019). Precipitation modification in cast Mg–1Nd–1Ce–Zr alloy by Zn addition. *Journal of Magnesium and Alloys*, 7(1), 113–123.

PART 2

Characterization

4

Preparation and Characterization of a Composite Material Using Sisal fibers for Light Body Vehicles

The present research work investigates the usage of local sisal plant leaves gathered from the Ethiopian highlands. The objectives of this chapter are to describe the processing techniques of specimen preparation and to examine the characteristics of sisal fibers as a composite material for light body vehicles on an experimental basis.

The fibers were pulled out through manual extraction with a knife. After extraction, the fibers were immersed in a solution containing 8 grams of sodium hydroxide (NaOH) in powder form and 10 liters of distilled water. The fibers were soaked for 24 hours to remove impurities and increase the surface roughness due to the disturbance of hydrogen bonding. The specimen was made using general-purpose resin (Gp) as a matrix and the hardener, and sisal fibers as a reinforcement material, which achieved the appropriate fiber orientation (0°, ± 45°, 0°, and 90°) by using an experimental (hand lay-up fabrication) technique and CATIA. The specimen was prepared according to the ECAE and ASTM standards.

Chapter written by Zewdie ALEMAYEHU, Ramesh Babu NALLAMOTHU, Mekonnen LIBEN, Seshu Kishan NALLAMOTHU and Anantha Kamal NALLAMOTHU.

4.1. Introduction

In order to protect natural resources and economize energy, weight reduction is the main focus of vehicle manufacturers at present. Weight reduction can be primarily accomplished by using better materials, design optimization and improving manufacturing processes. Even though there are several factors that influence the entire product development process of a lightweight vehicle, from the point of view of vehicle structural design, the major criteria for material choice are rigidity and strength properties that will determine the overall capability of the vehicle under static and dynamic loading conditions Koricho (2012, pp. 44–74).

Conventionally, the materials used in the construction of vehicle bodies are mostly different grades of steel. Although aluminum-intensive body concepts started being used in managerial class cars, before later being applied to other car classes, plastics mainly dominate the vehicle interior. Their external application is mostly limited to non-load-bearing components. Recently, some modern plastic materials have been approved for use in some automobile parts such as engine frame and forward bumper subsystems to decrease vehicle weight and achieve better occupant and pedestrian safety Cantor *et al.* (2008, pp. 3–27) and Koricho (2012, pp. 44–74).

4.1.1. *Statement of the problem*

Automobile manufacturing companies in Ethiopia are growing, and their main function is assembly of various kinds of vehicles. Examples are the Bishoftu Automotive and Locomotive Industry under the Metal and Engineering Corporation (METEC), Lifan Motors and Belayab Motors. The components to be assembled are not manufactured on site, they are imported from different car production enterprises outside Ethiopia, for example in China. However, some of the fiber composite body components are manufactured in Ethiopia by Dejen Aviation Industry (DAVI), but their inputs mainly depend on imported synthetic fibers for the fabrication of lightweight car body parts. However, those artificial fibers can lead to extra cost, the country's useful resources left unused and also increased release of environmental pollutants during disposal of the used composite material. This study aims to minimize the above problems by using a sisal fiber-reinforced unsaturated polyester resin composite material for the production of automobile body component parts locally in both small and large vehicle body maintenance garage, which contributes directly to the automotive industry and indirectly to Ethiopia's ability to generate or save hard currency and provides expanded opportunities for rural people as sisal fiber can be produced locally in Ethiopia.

4.1.2. *General objective*

The general objective of this research work is to prepare a lightweight composite material using sisal fiber for the construction of vehicle body panels.

4.1.3. *Specific objectives*

The specific objectives are as follows:

– to prepare a composite material using sisal fiber and unsaturated polyester resin;

– to characterize the mechanical properties of a fabricated composite material using tensile tests, compression, bending and water absorption tests, etc.;

– to model a body panel using CATIA software;

– to compare the properties of the prepared composite material with the properties obtained in previous work;

– to prepare a prototype of the body panel using a sisal fiber-reinforced composite.

4.1.4. *Significance of the study*

The application of a successful strategy to the manufacture of automobile bodies will render a lightweight but high-strength and rigid-body panel with enhanced fuel economy, reduced noise, high vibration absorption, durability, higher impact, fatigue and corrosion resistance. The composite material also costs less than the metallic one. There will be job opportunities for micro- and middle-level enterprises in the manufacturing areas. The reduced import of composite materials for automobile bodies ultimately reduces the cost.

Beneficiary: Automotive industries, vehicle owner, society, researcher and the country as a whole.

4.2. Literature review

4.2.1. *Introduction*

The literature review is focused on highlighting the importance of the present study in filling the existing research gap and to provide an overview of the work done to date towards the mechanical description of ordinary fiber composites and their properties. The most important topics included in this review are previous works on:

– natural fiber polymer composites;

– historical development of composite materials;

– overview of composites;

– natural fibers;

– test apparatus.

4.2.2. *Previous works on natural fiber polymer composites*

Guptaa and Srivastava (2016) discussed the properties of sisal fiber-reinforced epoxy composites. According to the literature, sisal fiber-reinforced epoxy composites have been organized by a hand lay-up approach after inactive compression, using different fiber weight divisions (15, 20, 25 and 30%). Mechanical properties, warm properties, moisture absorption properties and active mechanical investigation of sisal composites are investigated. The results show that the growth of sisal fiber in an epoxy framework produces up to 30 wt. % increments in the mechanical, thermal and water absorption properties. The highest values of tensile strength and tensile modulus obtained for the composite S30 are 83.96 ± 6.94 MPa and 1.58 ± 0.08 MPa respectively due to the strong grip between the sisal fibers and epoxy matrix, which permits a uniform exchange of stretch from the framework to the fibers.

Madhusudhana Reddy *et al.* (2014) carried out an experimental investigation of the mechanical properties of a sisal fiber and rice husk-reinforced polymer composite. They added 20, 15, 10 and 5 wt.% of sisal fiber to the resin at the most favorable percentage, then added 5, 3 and 1% of rice husk and replicated the to induce the crossover composite prepared using the hand lay-up procedure. They conducted the tests according to ASTM standards to find the tensile strength and flexural (bending) strength, which were 32.66 MPa and 57.99 MPa respectively.

Prasanna Venkatesh *et al.* (2016) found the tensile properties, flexural impact properties and moisture absorption properties of NFRP hybrid composites. They investigated the effects of fiber behavior and concentration on the mechanical properties of a short common fiber RPH composite. The fibers were treated with a 10% NaOH solution for 24 hrs. Fibers with distinctive lengths (5, 10 and 15 cm) and weight percentages (10, 15 and 20) were blended with unsaturated polyester for the initial planning of the composites. The mechanical properties concerning composites together with treated fibers were compared with UFC, and the mechanical and moisture absorption properties of treated/untreated hybrid composites were found to be 38.4 MPa/23.4 MPa (tensile strength), 78.2 MPa/56.7 (flexural strength) and 9.1/19.6 (water absorption).

4.3. Materials and methods

	Name	Quantity
1	Sisal Fiber	5 Kg
2	General Purpose Resin(GP)TOPAZ-1110TP	6 liter
3	Hardner (catalyst)	1 liter
4	Gelcoat White TOPAZ-1900W,G.P	2 liter
5	Wax	0.5 Kg
6	Roller brush	2 pc
7	Normal brush	4
8	Gloves	100 pc
9	Permanent Marker	5 pc
10	Scissors	1 pc
11	Vernior caliper	1 pc
12	Steel Rule	1 pc
13	Weighing machine	1 pc
14	Band saw	1 pc

Table 4.1. *Materials and supplies*

In this investigation, which aims to substitute mild steel with a composite material, several varieties of material and equipment are used for fabrication of composite materials and for testing the strategy applied to accomplish the objectives. The materials and their quantities used in this study are listed in Table 4.1.

4.3.1. *Sample preparation methods*

4.3.1.1. *Sisal fiber-reinforced composite preparation*

The guideline for the preparation of SFRUPR composite is shown in Figure 4.1.

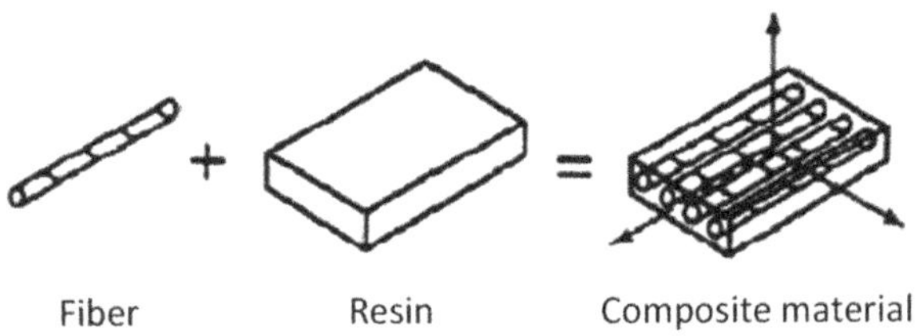

Figure 4.1. *Fabrication approach (3)*

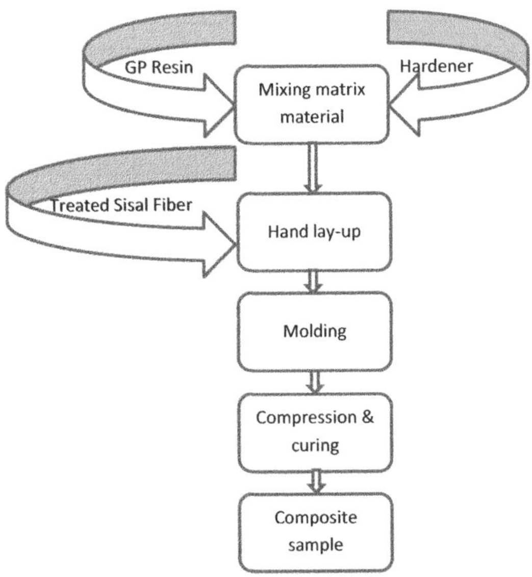

Figure 4.2. *Flow diagram for fabrication of sisal fiber-reinforced unsaturated polyester composites. For a color version of this figure, see www.iste.co.uk/kumar/materials.zip*

Figure 4.2 shows the scheme for the preparation of SFRUPR composites. The preparation of SFRUPR composites is accomplished by the hand lay-up method with sisal fiber with different orientations used as the reinforcement resin (general-purpose resin) and hardener used as the matrix material. The reinforcement is chosen based on its weight reduction, tensile strength and bending. In this research, 0, 45 and 90 degree orientations of sisal fiber were used.

Epoxy of general-purpose resin (GP) mixed with hardener is used to prepare the composite plate. The mass ratio of the mixture of resin and hardener is 100:1. Hardeners include acids, amines, polyamides, dicyandiamide, etc. The mixer is dispersed over with a stirrer for about 1 minute continuously. Mixing is performed in mixing containers (bowl) made of nickel, to prevent melting of the bowl due to the exothermic reaction, as shown in Figure 4.3.

Figure 4.3. *Preparation of unsaturated polyester resin (general purpose resin) and hardener. For a color version of this figure, see www.iste.co.uk/kumar/materials.zip*

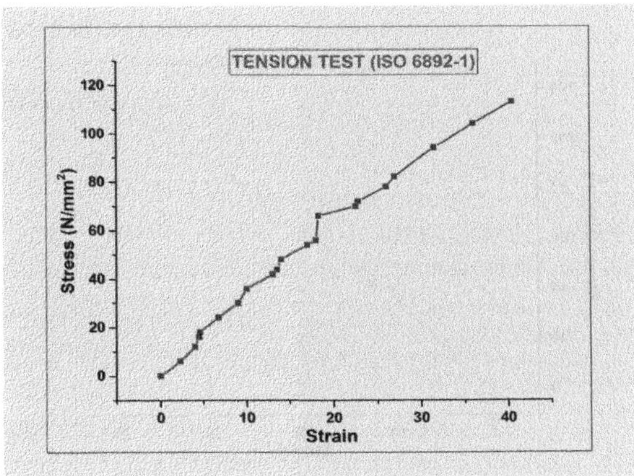

Figure 4.19(a). *T1(55 mm x 20 mm x 4 mm) – (0,90)° fiber orientation (T_{str} = 107 MPa). For a color version of this figure, see www.iste.co.uk/kumar/materials.zip*

Figure 4.19(a) presents the result of the tensile strength test at (0,90) degree fiber orientation. The percentage strain rate increases because of the forward elongation of toughness from 0% to 41.3%, and failure occurred after 41.3%.

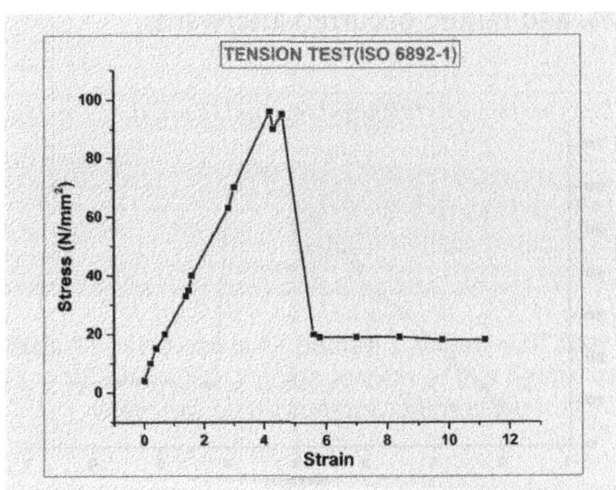

Figure 4.19(b). *T2 (60 mm x 21.3 x 4 mm) – (0,90)° fiber orientation (T_{str} = 87 MPa). For a color version of this figure, see www.iste.co.uk/kumar/materials.zip*

Figure 4.19(b) demonstrates the result of the tensile strength test at (0,90)° fiber orientation. The proportion of stress rate increases because of the forward elongation of durability from 0% to 4.8%, and failure occurred below 4.8%.

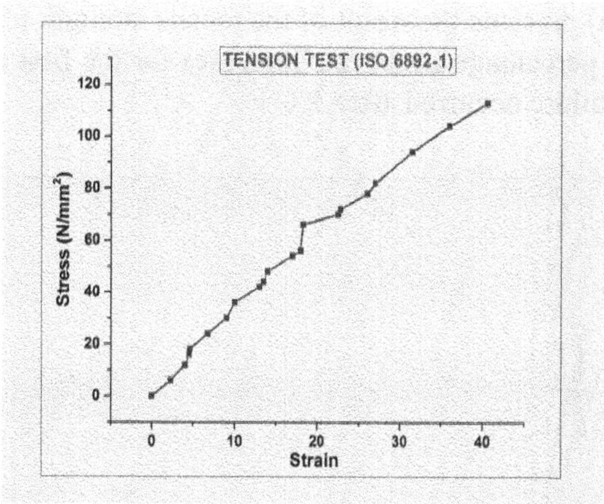

Figure 4.19(c). *T3 (60 mm x 20 mm x 4 mm) – (0,90)° fiber orientation (T_{str} = 107 MPa). (a–c). (0,90)° tensile strength test. For a color version of this figure, see www.iste.co.uk/kumar/materials.zip*

Figure 4.19(c) displays the result of the tensile strength test at (0,90)° fiber orientation. The percentage strain rate increases because of the forward elongation of stability from 0% to 19.1%, and failure occurred after 19.1%.

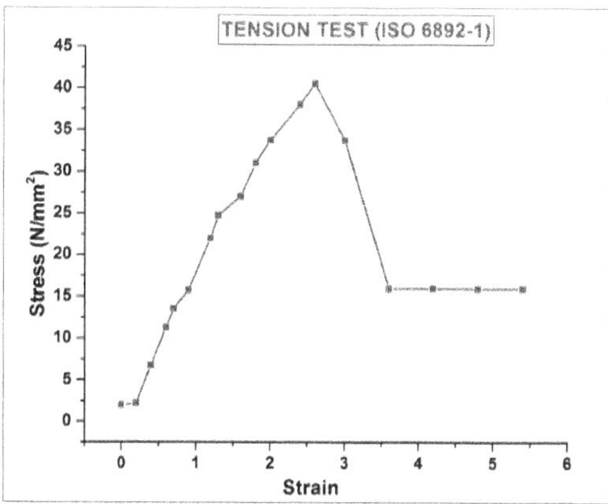

Figure 4.20(a). *T1 (60 mm x 20.5 mm x 4 mm) ± 45° fiber orientation (Tstr = 39 MPa). For a color version of this figure, see www.iste.co.uk/kumar/materials.zip*

Figure 4.21(b) shows compressive strength test at 0 degree fiber orientation; the compressive strength follows an increasing trend from 0% to 3.77%, failure occurred after 3.77%.

Figure 4.21(c) presents compressive strength test at 0 degree fiber orientation; the compressive strength follows an increasing trend from 0% to 4%, and failure occurred after 4%.

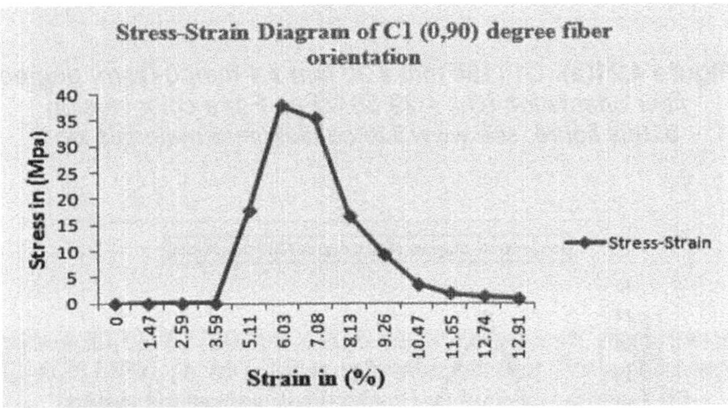

Figure 4.22(a). *C1 (155 mm x 30 mm x 4 mm) (0,90) degree fiber orientation (Cstr = 37.92 MPa). For a color version of this figure, see www.iste.co.uk/kumar/materials.zip*

Figure 4.22(a) shows compressive strength test at (0,90) degree fiber orientation; the compressive strength follows an increasing trend from 0% to 6.03%, and failure occurred after 6.03%.

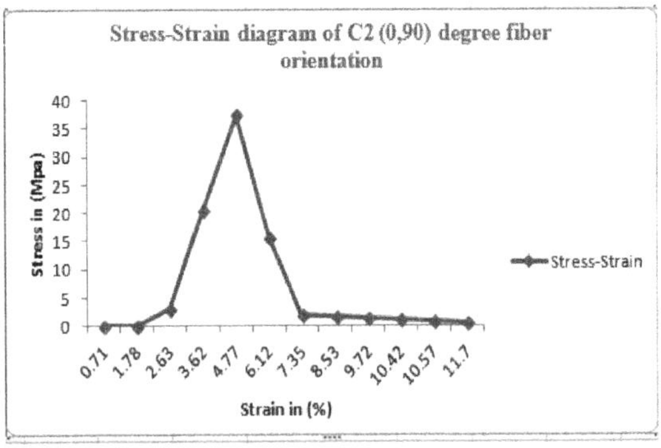

Figure 4.22(b). *C2 (155 mm x 30 mm x 4 mm) (0,90) degree fiber orientation (C_{str} = 37.42 MPa). For a color version of this figure, see www.iste.co.uk/kumar/materials.zip*

Figure 4.22(b) demonstrates compressive strength test at (0,90) degree fiber orientation. The compressive strength follows an increasing trend from 0% to 4.77%, and failure occurred after 4.77%.

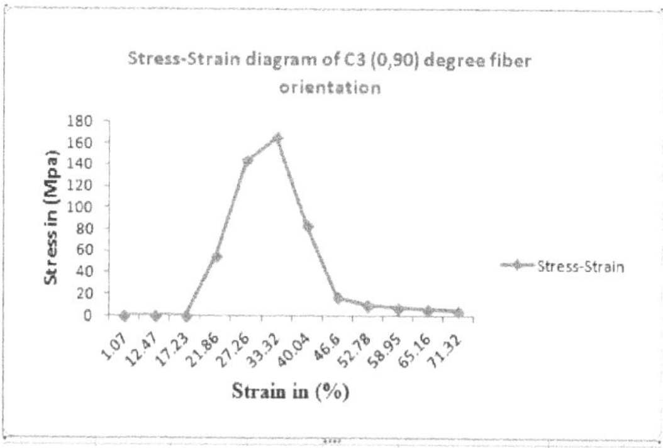

Figure 4.22(c). *C3 (155 mm x 30 mm x 4 mm) (0,90) degree fiber orientation (C_{str} = 165.52 MPa). (a–c) shows (0,90) degree compressive strength test. For a color version of this figure, see www.iste.co.uk/kumar/materials.zip*

Figure 4.22(c) presents compressive strength test at (0,90) degree fiber orientation. The compressive strength follows an increasing trend from 0% to 33.32%, and failure occurred after 33.32%.

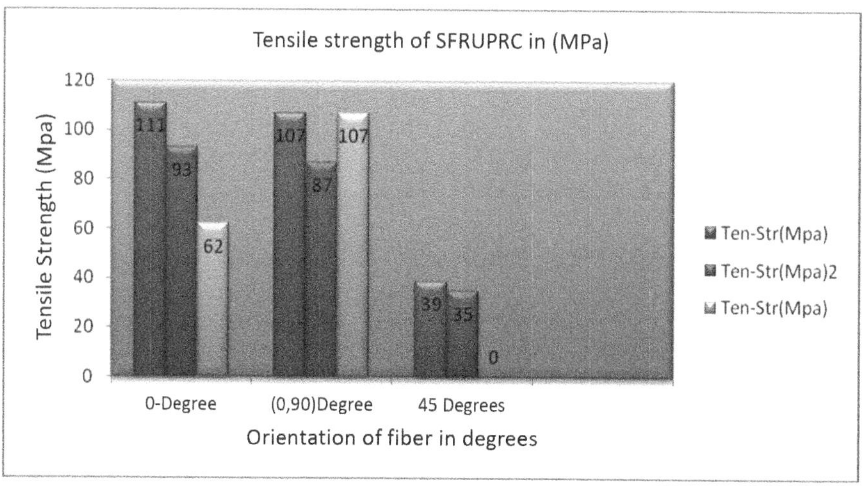

Figure 4.23. *Variation of tensile strength with fiber orientation For a color version of this figure, see www.iste.co.uk/kumar/materials.zip*

Preparation of fender prototype. The tensile strength is maximum in the case of (0,90)° fiber orientation. More load was required for rupture of laminates in the case of (0,90)° fiber orientation. Maximum elongation is observed at (0,90)° orientation.

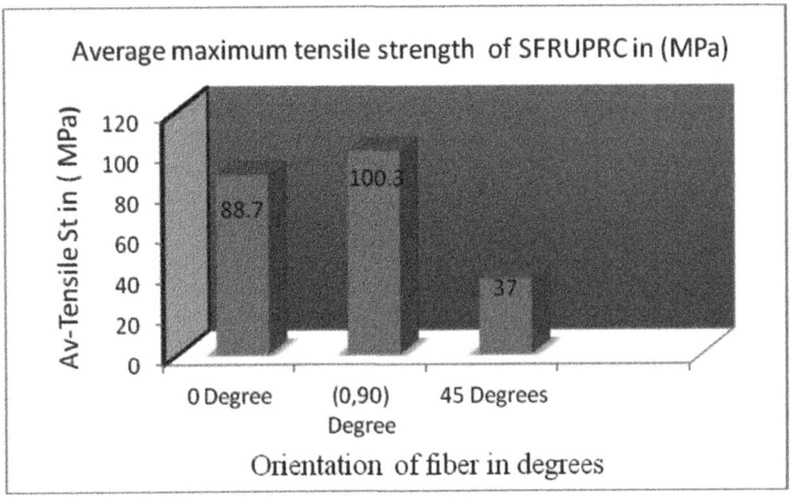

Figure 4.24. *Variation of average tensile strength with fiber orientation. For a color version of this figure, see www.iste.co.uk/kumar/materials.zip*

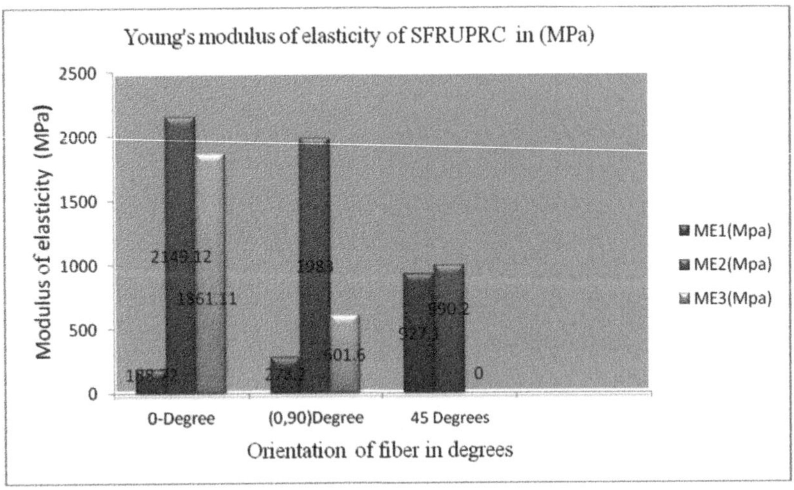

Figure 4.25. *Young's modulus of elasticity with different fiber orientations. For a color version of this figure, see www.iste.co.uk/kumar/materials.zip*

The figures show the tensile strength of SFRUPR composites at different orientations, among which the (0,90)° orientation shows the maximum tensile strength.

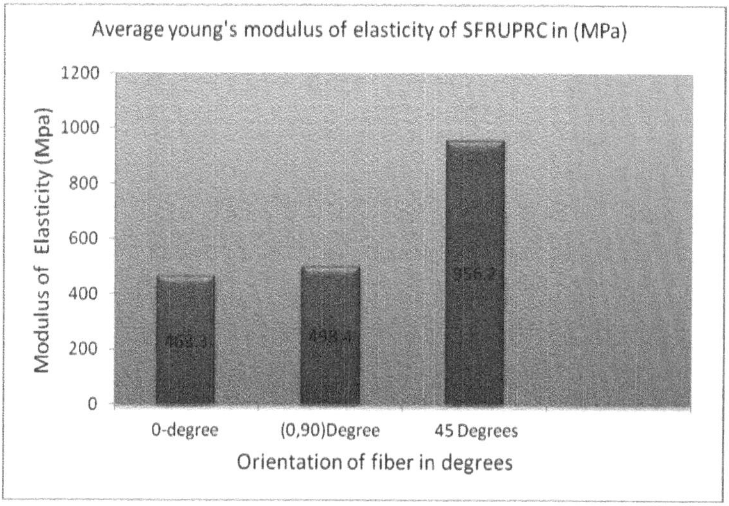

Figure 4.26. *Average Young's modulus with different fiber orientations. For a color version of this figure, see www.iste.co.uk/kumar/materials.zip*

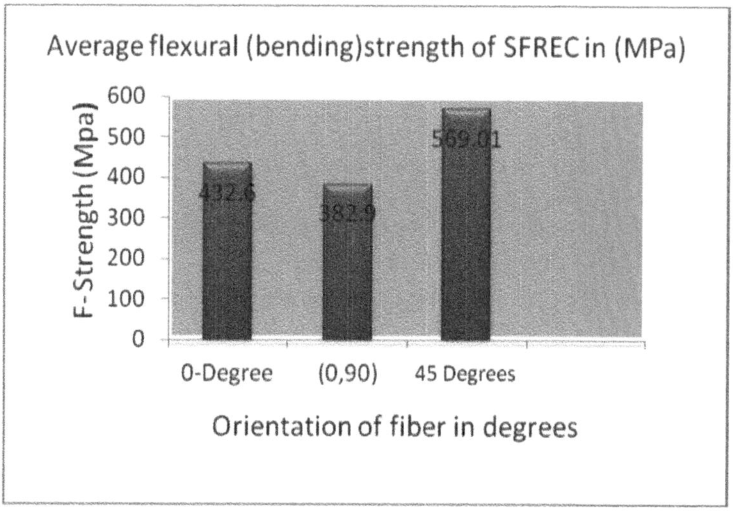

Figure 4.27. *Average flexural strength with different fiber orientation. For a color version of this figure, see www.iste.co.uk/kumar/materials.zip*

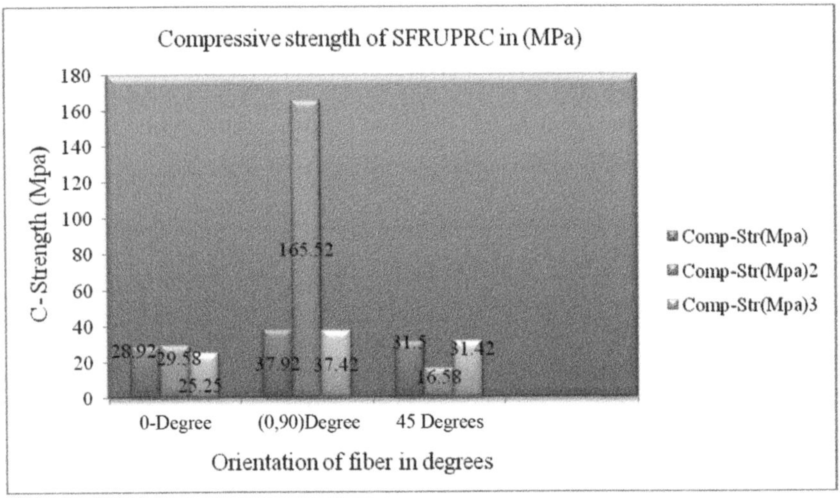

Figure 4.28. *Compressive strength with different fiber orientations. For a color version of this figure, see www.iste.co.uk/kumar/materials.zip*

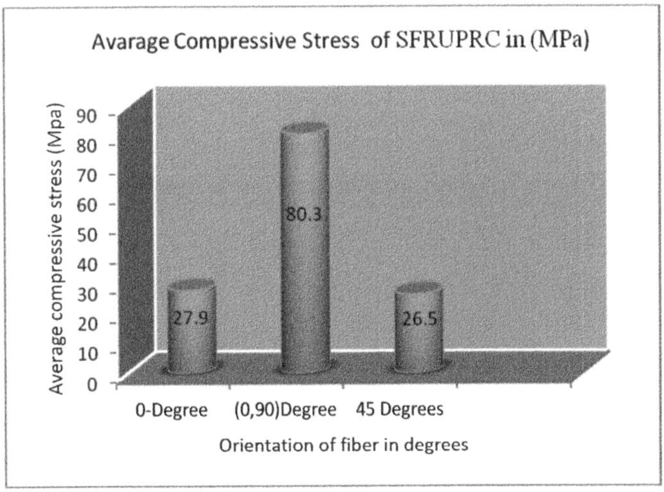

Figure 4.29. *Average compressive stress with different fiber orientations. For a color version of this figure, see www.iste.co.uk/kumar/materials.zip*

The figure shows the compressive strength of sisal fiber-reinforced unsaturated polyester resin composites at different orientations, among which the (0,90)° orientation shows the maximum compressive strength.

4.4.1.3. *Water absorption test*

The moisture absorption technique is used to determine the capacity of water absorption by a composite. It was observed that the percentage of water assimilation increased by increasing fiber content of all cases. The increase in water absorption is due to the hydrophilic nature of sisal fibers and more prominent interfacial region between the fibers and matrix. The weight was measured at an accuracy of 0.0001 g using four-digit equalizations (analytical precision balance) shown in the figure below.

Figure 4.30. *Analytical precision balance machine. For a color version of this figure, see www.iste.co.uk/kumar/materials.zip*

The following equation is used to determine water absorption:

$$\text{Water absorption} = \frac{W2 - W1}{\text{W1}} \text{ x } 100 \text{ in } (\%)$$

where W1 is the weight of the dry sample (weight before soaking into water) and W2 is the weight of the wet samples (weight after soaking into water (g)).

The test results, shown in Table 4.1 and the three pie charts in Figures 4.31–4.33, are based on the percentage of moisture absorption.

Orientation in (degree)	No of Sample	Thickness	Dry Samples (W1)	Wet Samples (W2)	W-Abs= $\frac{W2-W1}{W1}$ 100 in (%)
0-Degree	Ws1 Ws2 Ws3	4 4 4	5 5.64 5.44	5.28 5.89 5.78	5.6 4.43 6.25
Average			5.36	5.65	5.41
(0,90) Degree	Ws1 Ws2 Ws3	4 4 4	5.1 5.94 5.52	5.42 6.24 5.88	6.27 5.05 6.52
Average			5.52	5.85	5.98
± 45 Degrees	Ws1 Ws2 Ws3	4 4 4	7.46 6.71 6.71	7.7 7.06 6.96	3.22 5.21 3.72
Average			6.96	7.24	4.02

Table 4.2. *Test results based on the percentage of moisture absorption*

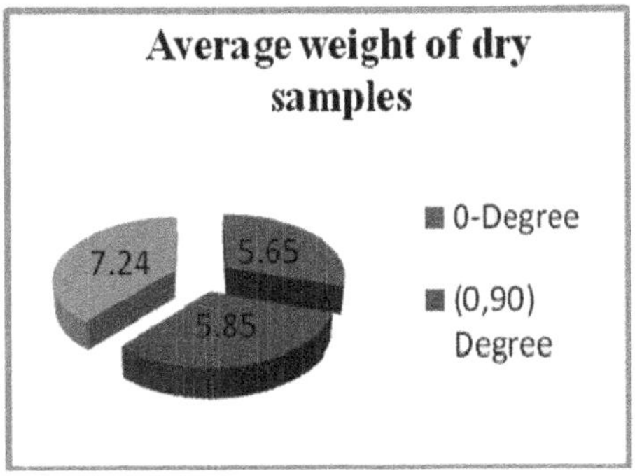

Figure 4.31. *Average weight of dry samples with different fiber orientations. For a color version of this figure, see www.iste.co.uk/kumar/materials.zip*

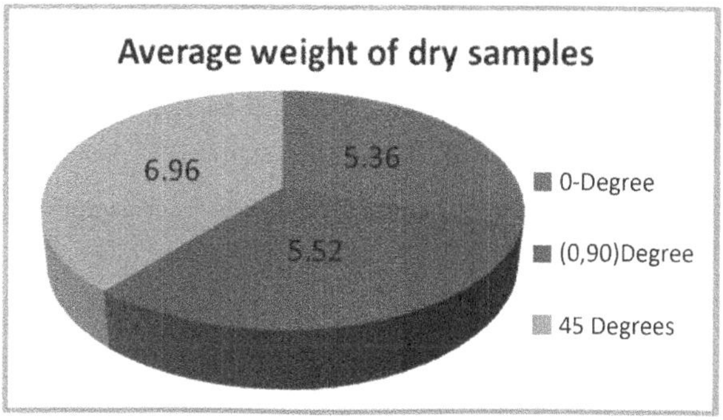

Figure 4.32. *Average weight of dry samples. For a color version of this figure, see www.iste.co.uk/kumar/materials.zip*

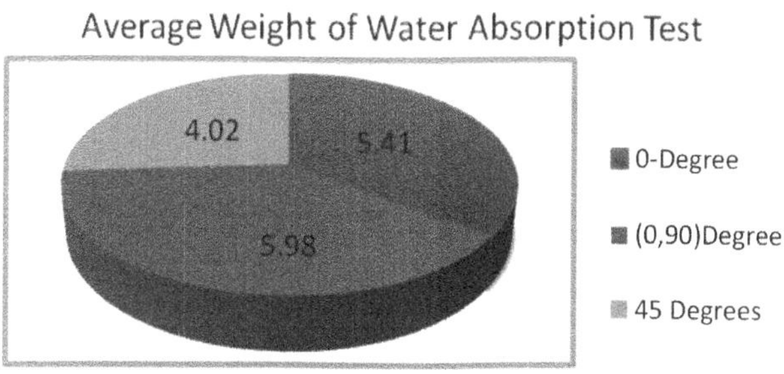

Figure 4.33. *Water absorption of different samples. For a color version of this figure, see www.iste.co.uk/kumar/materials.zip*

4.4.2. *Observation*

In these tests, all of the sisal fiber-reinforced unsaturated polyester resin composite specimen's fiber orientations have minimum water absorption, which is less than 15% of the ASTM C220 standards.

4.4.2.1. *Fender*

Fenders are complicated panels of sheet metal present on the sides cars, right by the door and over and around the front and rear wheels. They often take the shape of the wheels. Some fenders are included in the anatomy of the automobile, and some are standalone panels that can be bought and easily installed. The curve that forms a fender begins where the front and back doors end and runs ahead or backward to either the front or rear bumper and ends at either the hood or the lid of the trunk (see Figure 4.34 for drawing using CATIA).

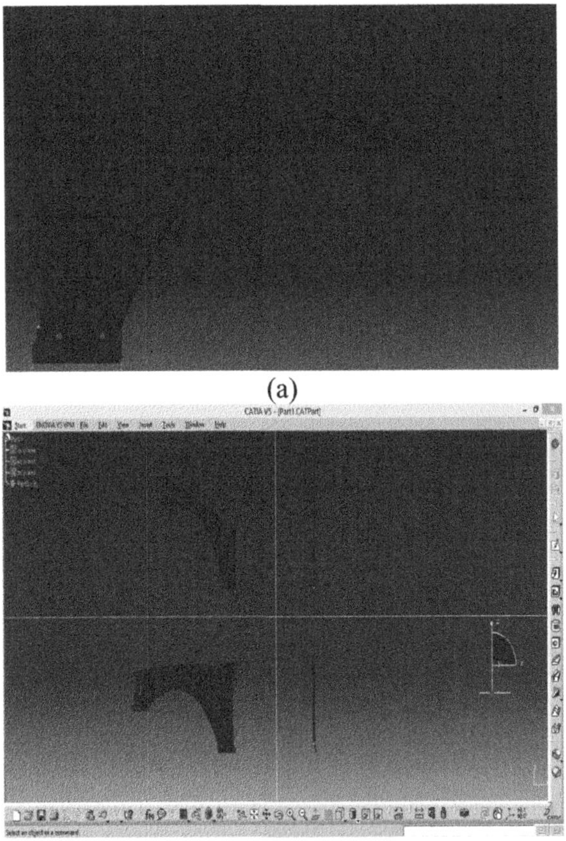

(b)

Figure 4.34. *Drafted model of fender using CATIA software: (a) all dimension in mm, (b) isometric view. For a color version of this figure, see www.iste.co.uk/kumar/materials.zip*

4.4.2.2. *Fender fabrication*

Based on the results obtained from the tensile, compressive, bending and water absorption tests conducted in mechanical and physical laboratories, it is concluded that it is favorable to manufacture vehicle body panels (fenders) that have relatively less stress, especially fender body panels locally at

0 (zero) and 90 degree fiber orientations using the hand lay-up procedure and make uniform dispersion of the matrix throughout the surface. The part is then cured at room temperature for 24 hours (Figure 4.35), and once solidified, it is deformed. The selected fiber orientation exhibited good mechanical properties.

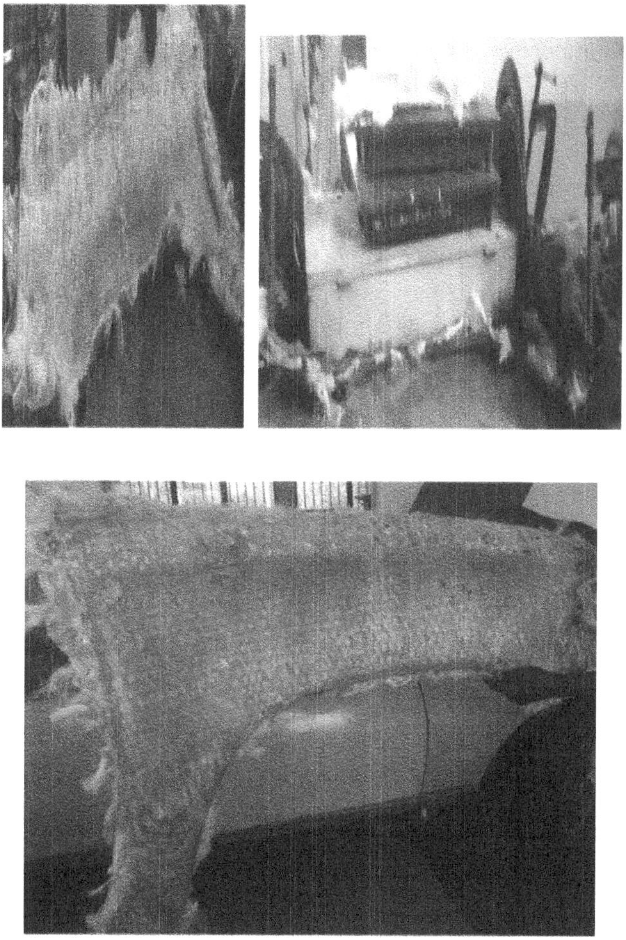

Figure 4.35. *Prototype of fender.*
For a color version of this figure, see
www.iste.co.uk/kumar/materials.zip

4.5. Comparison of previous works

4.5.1. *Tensile strength*

Most of the works concerning the tensile properties of natural fiber RUPR composites are compared with the current work. From Table 4.3, the tensile strength found in the current study is not similar to that found in the literature. Composites had been organized using various fiber weight fractions (15, 20, 25 and 30 wt.%) with unidirectional alignment of sisal fiber epoxy resin for the initial preparation of the composites. It can be concluded that good results are obtained from the current work. Fiber orientation in degrees is important to a SFRUPR composite feature if the sisal fiber is (0,90) degrees fiber tensile properties are high comparing 0 degree and ± 45 degree fiber orientation. Therefore, SFRUPRC of (0,90) degree fiber orientation has great potential for the automotive body application of fiber-reinforced polymer composites.

Matrix material	Fiber	Average Max-Tensile Strength(Mpa)	Reference	Author
Epoxy + Hardener	Sisal	83.96 ± 6.94	Previous work	M K Guptaa & R K Srivastava
unsaturated polyester resin+catalyst(hardener)	Sisal/Bamboo 25 % / 75 %	23.5	Previous work	R. Prasanna Venkatesh1, K. Ramanathan2,
Unsaturated polyester resin + Hardner	Rice husk and Sisal Fiber 20 % / 80 %	32.66	Previous work	Madhusudhana Reddy H1, Bharath S Kodli2, Ravi B Chikmeti
Unsaturated Polyester Resin+ Hardener	100 % Sisal Fiber	100.3	Current Work	Current Researcher

Table 4.3. *Comparison with previous works on the tensile properties of sisal fiber composites*

4.5.2. *Bending strength*

Some other studies regarding bending strength of natural fiber composites are also compared with the current work. From Table 4.4, the bending strength found in the current work is higher than those found in previous works.

The current price of vehicle fenders in Ethiopia is 8,000–12,000 birr from second hand (from Dubai) to the original genuine part (from MOENCO). However, manufacturing such a fender with the same shape may cost only 3,500 birr, including all the necessary composite materials and technician costs, but the thickness and molding are different from the steel or plastic front and rear bumpers.

Matrix material	Fiber	Average Max-Bending Strength(Mpa)	Reference	Author
Epoxy + Hardener	Sisal	252.39□12.11	Previous work	M K Guptaa & R K Srivastava
unsaturated polyester resin+catalyst(hardener)	Sisal/bamboo 25 %/75 %	57.8	Previous work	R. Prasanna Venkatesh1, K. Ramanathan2,
Unsaturated polyester resin + Hardner	Rice husk and Sisal Fiber 20 %/80 %	57.99	Previous work	R.Sakthivela*, D.Rajendranb
Unsaturated Polyester Resin+ Hardener	100 % Sisal Fiber	569.01	Current Work	Current Researcher

Table 4.4. *Comparison with previous works on bending properties of sisal fiber composites*

4.6. Conclusion and recommendation

Based on the tensile, compression, flexural and water absorption properties obtained from the experimental data in this work, a few points can be summarized as follows:

– From the experiment result, it was observed that different orientations of fiber showed enhanced mechanical properties of the sisal fiber composite material.

– From the test results of tensile strength, it was found that 0° and 90° (bidirectional)-treated sisal fiber-reinforced unsaturated polyester resin composites have better tensile strength than the 0° and ± 45° ones.

– From the test results of flexural strength, it was found that ± 45°-treated sisal fiber-reinforced unsaturated polyester resin composites have better flexural strength than the 0° and ± 45° ones.

– The effect of moisture content obtained from the water absorption test is within the specifications and has minimum water absorption for auto body panels.

– From the water absorption test results of each specimen, the ± 45 degree fiber orientation shows 4.02%, which is lower than the other two fiber arrangements (5.41% (0 degree) and 5.98% (0,90) degree). In these tests, all of the SFRUPRC specimen's fiber orientations have minimum water absorption, which is less than 15% of the ASTM C220 standards.

4.6.1. *Recommendations for prospective applications*

Sisal fiber reinforced-unsaturated polyester resin composite materials are appropriate for structural automotive applications, such as interior and exterior body panels, and have comparable standards to steel components.

Sisal fiber-reinforced unsaturated polyester resin composites were fabricated and evaluated for their mechanical and physical characteristics. This investigation found that there are sufficient potential applications in the following fields: manufacturing of lightweight sports goods such as tennis rackets, balls, bicycle frames, snowboards, spur and helical gears.

The present study found that sisal fiber-reinforced unsaturated polyester resin composites are used when cost reduction is the prime objective, and can effectively replace the conventional and relatively expensive materials. Their use may be recommended for household applications such as the fabrication of tables, chairs, door panels, interior paneling, door-frame profiles, food trays, partitions, bath units, lampshades, suitcases and helmets.

Production of some other vehicle structures may also use composite materials to show comparable advantages to the automobile, educational and government fields.

4.6.2. *Scope for future work*

The necessity for continuing this work is presented in this section. The following areas are suggested for further study:

– impact, fatigue and shear test behavior of sisal fiber-reinforced unsaturated polyester resin composites;

– modification of the mechanical properties of SFRUPR by adding nano-clays;

– scanning electron microscopy (SEM) and finite element analysis.

Tribological assessment of sisal fiber-reinforced unsaturated polyester resin composite has been a much less studied area. There is a very wide scope for future scholars to explore this area of research.

4.7. References

Cantor, B., Grant, P., and Johnston, C. (2008). *Automotive Engineering Lightweight, Functional, and Novel Materials*. Taylor & Francis Group, Boca Raton, FL, USA.

Cholachagudda, V.V., Udayakumar, P.A., and Ramalingaiah (2013). Mechanical characterisation of coir and rice husk reinforced hybrid polymer composite. *International Journal of Innovative Research in Science, Engineering and Technology*, 2(8), 3779–3785.

Guptaa, M.K. and Srivastava, R.K. (2016). Properties of sisal fibre reinforced epoxy composite. *Indian Journal of Fibre & Textile Research*, 41, 235–241.

Hardinnawirda, K. and Sitirabiatull Aisha, I. (2012). Effect of rice husks as filler in polymer matrix composites. *Journal of Mechanical Engineering and Sciences (JMES)*, 2, 181–186.

Chand Badshah S.B.V.J., K. Arun, B. Eswaraiah (2015). Fabrication and testing of natural (sisal) fiber reinforced polymer composites materials. *International Journal of Emerging Trends in Engineering Research*, 3(5).

Koricho, E.G. (2012). Implementation of composite and plastics materials for vehicle light weight. PhD thesis, Polytechnic of Turin, Italy.

Madhusudhana Reddy, H., Kodli, B.S., and Chikmeti, R.B. (2014). Experimental investigation of mechanical properties of sisal fiber and rice husk reinforced polymer composite. *IOSR Journal of Mechanical and Civil Engineering*, 11(4), Ver. V, 05–11.

Malhotra, N., Sheikh, K., and Rani, S. (2012). A review on mechanical characterization of natural fiber reinforced polymer composites – Review article. *Journal of Engineering Research and Studies*, III(I), 75–80.

Negera, D.W., Bhaskaran, J., Ilmi, I., and Nallamothu, R.B. (2019). Characterization of hybrid composite made of false banana fiber and sisal fiber. *International Journal of Engineering and Advanced Technology (IJEAT)*, 9(2).

Prasanna Venkatesh, R., Ramanathan, K., and Srinivasa Raman, V. (2016). Tensile, flexual, impact and water absorption properties of natural fibre reinforced polyester hybrid composites. *Fibres & Textiles in Eastern Europe*, 24(3), 90–94. DOI: 10.5604/12303666.1196617.

Sakthivel, M. and Ramesh, S. (2013). Mechanical properties of natural fiber (banana, coir, sisal). *Polymer Composite*, 1(1).

Shin, K.C., Lee, J.J., Kim, K.H., Song, M.C., and Huh, J.S. (2002). Axial crush and bending collapse of an aluminum/GFRP hybrid square tube and its energy absorption capability. *Composite Structures*, 57(1–4), 279–287.

Verma, D., Gope, P.C., Shandilya, A., Gupta, A., and Maheshwari, M.K. (2013). Coir fibre reinforcement and application in polymer composites: A review. *Journal of Materials and Environmental Science*, 4(2), 263–276.

5

Optimizing the Polystyrene Catalytic Cracking Process Using Response Surface Methodology

The statistical response surface method was used to optimize the experimental design of the catalytic cracking process of plastic waste. The C:F ratio, operating temperature and time requirement have been chosen as effective parameters of the experiments. The Box–Behnken design has been used to develop a second-order quadratic model. The optimum process conditions of the design are 0.25 C:F ratio, 450°C reaction temperature and 45 min time. These resulted in a higher fraction of liquid yield of 88.7%. In addition, experimental investigation of the polystyrene catalytic cracking process was conducted by using a calcinated silica-magnesium catalyst. As an outcome of the experiments, the optimal conditions are 1:4 C:F ratio, 450°C reaction temperature and 45 min reaction time, which resulted in a higher liquid fraction of 91.2%. These similarities provided a reasonable agreement between experimental and statistical operating conditions. The characterization of the obtained liquid product was performed using FTIR spectroscopy, GC/MS analysis and physico-chemical analysis. The GC/MS analysis of the obtained liquid yield shows that the silica-magnesium cracking of polystyrene leads to the production of gasoline-range products within C_3–C_{13}. The FTIR analysis shows that the appearing hydrocarbon consists of aliphatic and olefinic compounds with a major constituent of

Chapter written by Selvaganapathy THAMBIYAPILLAI, Muthuvelayudham RAMANUJAM and Jayakumar MANI.

aromatic compounds. Hence, it can be concluded that the developed quadratic model was well fit to the experimental domains and the predicted operating conditions are most suitable for conducting catalytic cracking experiments.

5.1. Introduction

Modern societies face challenges towards increasing research activities of plastic waste management worldwide. Currently, plastic waste is considered as an international agenda for both industrial and municipal solid wastes (Amoodi *et al.* 2013; Awasthi *et al.* 2017). Plastic wastes in both streams are mainly composed of polyethylene, polyvinyl chloride, polyethylene terephthalate, polypropylene and polystyrene (Kunwar *et al.* 2016). The generation of plastic waste in municipal plastic waste has reached about 150 metric tons/annum (TPA) all inclusive. In India, the generation of plastic wastes during 2018–2019 is approximately 3,360,043 tons/annum (TPA) (Ministry of Housing and Urban Affairs 2019). About 50–70% of plastic wastes are packaging items mostly containing recyclable plastics such as polyethylene, polypropylene, polystyrene and polyvinyl chloride. Polystyrene is the main component of municipal plastic waste, which contributes 10% of total plastic wastes. Used polystyrene plastics from food packaging, laboratory ware, electronic households and single-use cups have led to serious environmental issues (Selvaganapathy *et al.* 2020). To overcome these issues, modern research works have focused on finding a safe disposal for plastic waste. According to a statistical report, about 10% of plastic waste was recycled, 25% was incinerated and the remaining 65% was moved to landfill sites (Vas *et al.* 2017). Although landfilling and thermal incineration are the most common methods, they are very costly and may produce undesirable emissions (Nisar *et al.* 2019).

Chemical recycling is an attractive method, which is generally applied to convert plastic waste into value-added hydrocarbon products. Thermal cracking and catalytic cracking are two approaches in the chemical recycling process (Shah *et al.* 2020). The requirement of a very high temperature, typically about 900°C, is the main drawback of the thermal cracking process. Catalytic cracking, which requires the presence of catalysts, is demonstrated as a significant conversion technique of plastic waste into useful fuel products at moderate temperatures and short retention times compared to thermal cracking (Patni *et al.* 2013; Alla and Ali 2014; Jouhara *et al.* 2018).

Silica alumina-based catalysts, FCC, bentonite, dolomite, zeolites and reforming catalysts were extensively applied in the catalytic cracking process of plastic waste (Miandad *et al.* 2016).

The yield distribution of the catalytic cracking process mainly relies on operating variables such as catalyst type, catalyst-to-feedstock ratio, temperature, relative pressure, time requirement and size of plastics (Istadi *et al.* 2010). Kumar and Singh (2014) carried out an optimization study of catalytic cracking of HDPE using the RSM. The CCD matrix was applied to optimize operating variables such as temperature, acidity of the catalyst and C:F ratio to maximize liquid product yield (Kumar and Singh 2014). Selvaganapathy *et al.* (2020) performed an optimization study of thermal cracking of polystyrene using the RSM.

The Box–Behnken design was applied to optimize the operating variables such as size and weight of feedstock, reaction time and reaction temperature to maximize liquid yield. Therefore, RSM is a statistical technique that can be extensively used to optimize the experimental design, that is, the Box–Behnken design (BBD). RSM involves three main aspects: optimizing, designing and modeling of experiments by selecting appropriate experimental domains (Cornell 1990; Miranda *et al.* 2010; Pinto *et al.* 2013). The objectives of the RSM strategies are to (a) design several experiments to achieve adequate outcomes of the response, (b) establish a Box–Behnken design (BBD) having good fit with the experimental domains, (c) determine the optimum conditions of the experiments and (d) reduce the number of experiments by choosing suitable parameters (Mahalik *et al.* 2010; Krishankant *et al.* 2012; Montgomery 2012).

This work investigates the process parameter optimization on the catalytic cracking process of plastic waste by applying the response surface method (RSM) using Design-Expert software; the developed design is called the Box–Behnken design. The three effective parameters, C:F ratio, operating temperature and time, are chosen in this work. In addition, an experimental investigation on the catalytic cracking process of polystyrene waste was conducted in a bench-scale pyrolytic reactor at temperatures ranging between 400 and 500°C. Furthermore, silica-magnesium-based heterogeneous catalyst was synthesized and used directly with different ratios of polystyrene such as 1:4, 1:2, 3:4 and 1:1 respectively. The liquid fuel derivative was analyzed by FTIR spectroscopy, GC/MS analysis, and physico-chemical analysis respectively.

5.2. Material and methods

5.2.1. *Materials*

Various types of plastic wastes were obtained from the waste dumping section of the Tiruvannamalai municipality, Tamilnadu, India. These were segregated into different plastic groups according to the Society of the Plastics Industry (SPI) guidelines. Polystyrene waste was selected and initially shredded into a small size of about 1 cm^2. The shredded plastic sample (PS) was then cleaned properly to remove unwanted waste.

Thermal analysis of polystyrene by thermogravimetric analysis (TGA) was carried out using a Universal V4 7A TA instrument under nitrogen (N_2) atmosphere. Precisely weighed, 3.238 mg of polystyrene was poured into a platinum HT pan and heated from 50 to 600°C at 10°C/min.

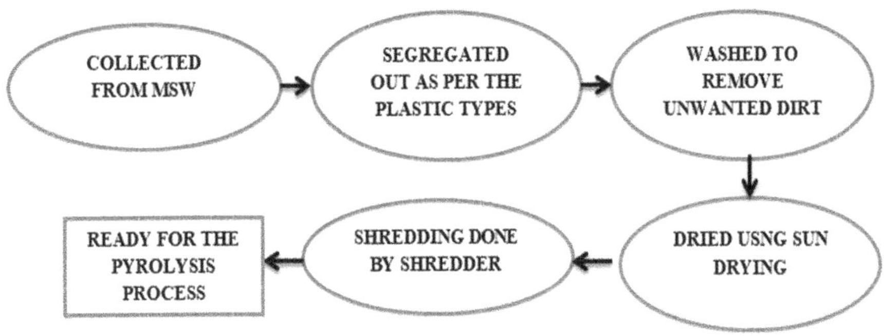

Figure 5.1. *Schematic illustration of preprocessing of polystyrene. For a color version of this figure, see www.iste.co.uk/kumar/materials.zip*

Figure 5.2. *Shredded waste polystyrene with its recycling code 6. For a colorversion of this figure, see www.iste.co.uk/kumar/materials.zip*

The mixed fraction of SiO_2-MgO material obtained from a local market was used as a heterogeneous catalyst for this investigation. Before the use of a catalyst, the raw SiO_2-MgO was synthesized by the calcination method to purposely increase its pore sizes, surface area and catalytic activity. About 10 g of the sample was finely crushed into a small size of about 50 μm by a roll crusher. After that, the fine powder was left to stand in a hot-air oven at 950°C for 3 hrs. After the standing period, the silica-magnesium was taken out from the oven and cooled down to ambient temperature in a desiccator and transferred to glass jars for experimental investigations.

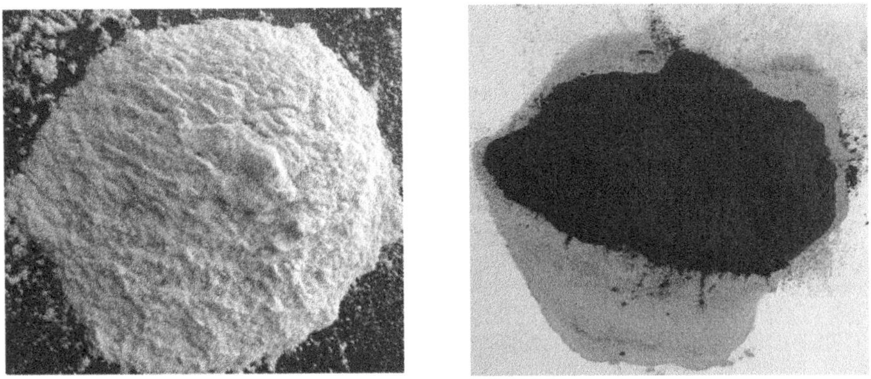

(a) (b)

Figure 5.3. *a) Raw SiO_2-MgO and b) 950°C calcined silica-magnesium catalyst. For a color version of this figure, see www.iste.co.uk/kumar/materials.zip*

The silica-magnesium catalyst was characterized by a simultaneous SEM-EDS analysis using a Hitachi X-650 analyzer equipped with a Quantax 200-Bruker device. The morphological structure of the catalytic material was determined using an SEM analyzer, and the atomic composition of the silica-magnesium catalyst was found by energy-dispersive spectroscopy (EDS).

5.2.2. *Experimental procedure and characterization*

The process involved in the catalysis of polystyrene is investigated in a bench-scale pyrolytic reactor, as explained in our previous study. In this study, the construction and working principles are elaborated. The experimental setup for the catalysis of polystyrene is shown in Figure 5.4.

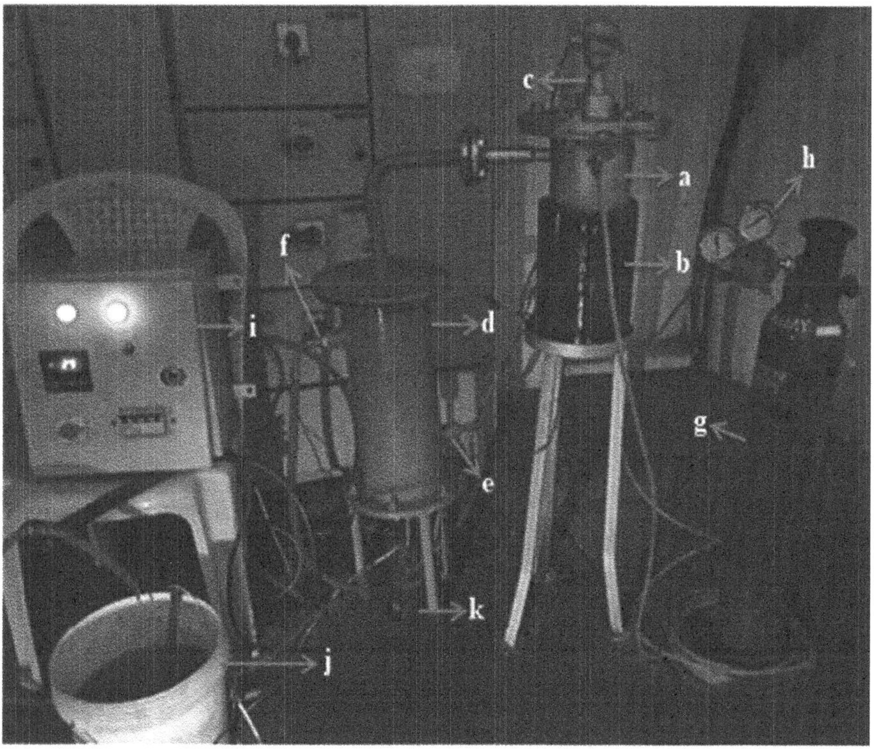

Figure 5.4. *Experimental setup of the pyrolysis process. (a – SS bench-scale pyrolytic reactor; b – band heater; c – temperature sensor; d – shell and tube condenser; e – cooling water in from water recirculation tank; f – cooling water out to water recirculation tank; g – nitrogen gas cylinder (purging gas); h – flow control valve of N2 cylinder; i – PID controller panel board; j – water recirculation tank; k – liquid product collection point). For a color version of this figure, see www.iste.co.uk/kumar/materials.zip*

A bench-scale pyrolytic reactor with 2 kg capacity was used, which withstands temperatures up to 1,000°C using a band heater that was controlled and sensed by a PID controller and a temperature sensor respectively. Purging nitrogen gas in the reactor prevents even the trace amounts of harmful contaminants or gases. A shell and tube condenser made of stainless steel with copper tubing was fixed at the reactor's outlet. The organic vapor produced from the reacting chamber was then condensed into the pyrolysis liquid oil using the shell and tube condenser with provisions for recirculated water. Before starting the catalytic cracking experiment, typically about 1:1, 1:2, 1:4 and 3:4 ratios of polystyrene along with silica-magnesium catalyst were loaded into the reacting chamber, which was heated at temperatures between 400 and 500°C with an increment of 10°C. The experimental investigation was initiated by loading the plastic with a catalyst in the reacting chamber and setting the temperature controller to the desired temperature level by turning the electric heater. After the experiment, the vapor left in the reactor was then condensed through a condenser. The operating time of every trial was monitored by the primary drop of the liquid–vapor produced. The bottom product such as char residue was taken out from the reacting chamber after completion of the experiment. The weight of pyrolysis products could be measured by the standard mass balance analysis.

The liquid fuel was characterized by FT-IR spectroscopy and GC/MS analysis to observe the functional group and the presence of hydrocarbon compounds in the obtained liquid fuel. In order to obtain the functional group, the typical molecular structure of the liquid fuel was analyzed within the wavenumber between 500 and 3,500 cm^{-1} and resolution number 5 cm^{-1}. GC/MS analysis was used to find the presence of hydrocarbon compounds in the obtained liquid product. The physico-chemical properties of the obtained liquid fuel were analyzed by ASTM standards.

5.2.3. *Design of catalytic cracking experiment using response surface methodology*

In this research work, the process parameters used for catalytic cracking of polystyrene were determined using the RSM. Optimization of the catalytic cracking experiment by RSM involves three stages: designing experiments, followed by estimating the coefficients of the quadratic equation and predicting the responses of liquid fuel. Therefore, the RSM is a statistical method that was the most suitable for experiment to achieve an optimum and adequate outcome of the responses. In this optimization study, three process parameters, namely (A) C:F ratio, (B) operating temperature and (C) time needed to complete an experiment, were selected for the optimization study, and each independent variable was assessed in 17 unique experiments.

The RSM was used to optimize the catalytic cracking experiment, which can be represented by equation [5.1]:

$$Y = f(A, B, C) \qquad [5.1]$$

where Y is the liquid fuel response and A, B, and C are the operating variables. The primary goal of this study is to design an experiment and to optimize the response of liquid yield (Y). The second-order polynomial equation for the prediction of liquid yield response is given in equation [5.2]:

$$Y = \beta_0 + \sum_{i=1}^{n} \beta_i X_i + \sum_{i=1}^{n} \cdot \beta_{ii} X_i^2 + \sum_{i=1}^{n} \cdot \sum_{j=i+1}^{n} \cdot \beta_{ij} X_i X_j + \varepsilon \qquad [5.2]$$

where Y is the liquid yield response; β_0, β_i, β_{ii} and β_{ij} are regression coefficients and n is the number of experiments.

The total number of experiments (N) was measured by equation [5.3]:

$$N = = 2^n + 2n + m = 2^3 + 2(3) + 3 = 17 \qquad [5.3]$$

Predicted values were obtained using equation [5.2] for each of the responses. These predicted values were obtained by plotting surface and contour plots, which are used to study the interaction effects between each parameter. Both plots were used to facilitate a mutual relationship between the process parameters that were made by setting both factors constant at their central axis. The adequacy of the model was checked by analysis of variance (ANOVA).

5.3. Results and discussion

5.3.1. *Thermal analysis of polystyrene*

The thermal analysis of waste polystyrene and the thermal curves are shown in Figure 5.5. The thermal analysis of polystyrene shows only one step of degradation; hence, the degradation of polystyrene is termed as "one-step degradation mechanism". The degradation of waste polystyrene began at 318.5°C and ended at 410°C, reaching 91.61% of degradation at the heating rate 10°C/min. Also, it was concluded that 50% degradation for PS was achieved around 425°C. J. Nisar *et al.* (2019) studied the thermal analysis of polystyrene at different heating rates and reported a temperature increase from 387 to 428°C. This can be attributed to the high degradation of polystyrene (Nisar *et al.* 2019).

5.3.2. *SEM-EDX analysis*

The pre-treatment of the silica-magnesium catalyst was established to enhance the catalytic performance such as pore size, surface area and catalytic activity of the catalyst. The objectives of the synthesis of the SiO_2-MgO material in an inert atmosphere are to remove any undesired species from the surface and to increase the liquid product yield by reducing the gas yield in catalytic cracking of waste polystyrene. The performance of the calcinated catalyst was measured using simultaneous SEM-EDS analysis. The morphological structure of the silica-magnesium catalyst was monitored by scanning electron microscopy (SEM), and its composition was calculated by energy-dispersive spectroscopy (EDS) analysis.

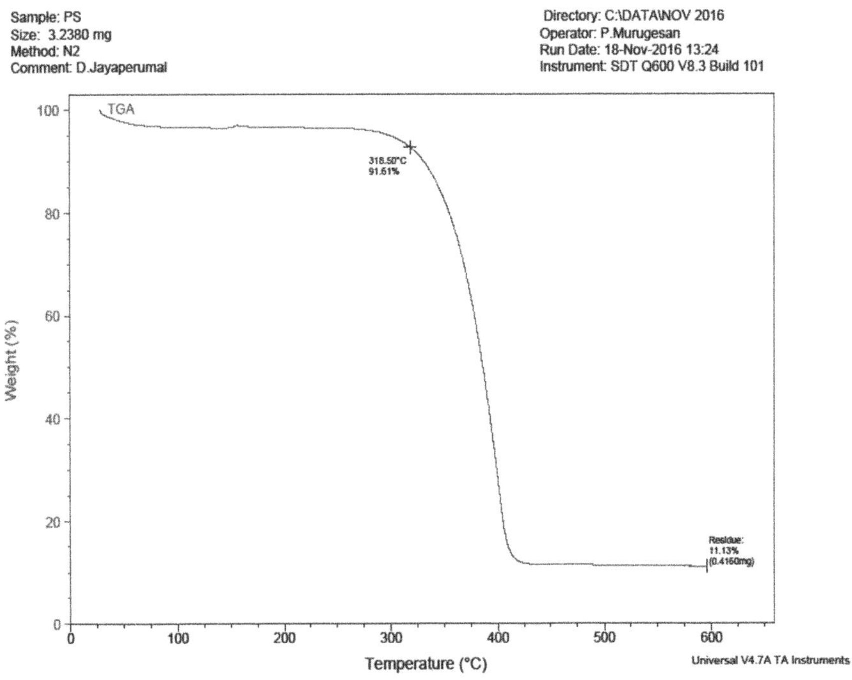

Figure 5.5. *Thermal degradation of polystyrene obtained from TGA analysis at 10°C/min. For a color version of this figure, see www.iste.co.uk/kumar/materials.zip*

The morphological structures of the raw catalyst and the calcinated catalyst are shown in Figures 5.6 and 5.7 respectively. The SEM image of raw SiO_2-MgO was observed to be smooth and contain larger grains with regular morphological structures. The morphological analysis of the calcinated silica-magnesium catalyst showed an undefined morphological shape that would have resulted due to the change in structure. The calcination effect of the silica-magnesium catalyst reveals that the catalytic activity of the catalyst was successfully enhanced, and this catalyst is recommended for the catalytic cracking process. The SEM analysis of a calcinated dolomite catalyst (Olivia *et al.* 2017) and impregnated fly ash catalyst (Babajide *et al.* 2010) was studied, revealing that the pore size of the catalyst increased with increasing catalytic activity. Both investigations are in good association with this study.

Figure 5.6. *Scanning of raw SiO_2-MgO material with 5 µm magnifications*

Figure 5.7. *Scanning of 950°C calcinated silica-magnesium catalyst with 5 µm magnification*

The surface atomic composition of raw SiO_2-MgO and calcinated silica-magnesium catalyst are listed in Table 5.1.

Elements	Weight %		Atomic %	
	Raw SiO_2-MgO	**950°C Calcinated silica-magnesium**	**Raw SiO_2-MgO**	**950°C Calcinated silica-magnesium**
O	55.18	54.11	67.08	66.13
Mg	17.50	17.76	14.00	14.28
Si	27.32	28.13	18.92	19.59

Table 5.1. *Atomic composition of raw SiO_2-MgO versus 950°C calcinated silica-magnesium catalyst*

As can be seen in Table 5.1, the weight percentages of raw SiO_2-MgO were found to be O (55.18%), Mg (17.50%) and Si (27.32%), with atomic composition of 67.08% O, 14.00% Mg and 18.92% Si respectively. After the calcination process, the weight percentages of the calcinated silica-magnesium catalyst were found to be O (54.11%), Mg (14.28%) and Si (19.59%) with atomic composition of 66.13% O, 14.28% Mg and 19.59% Si respectively.

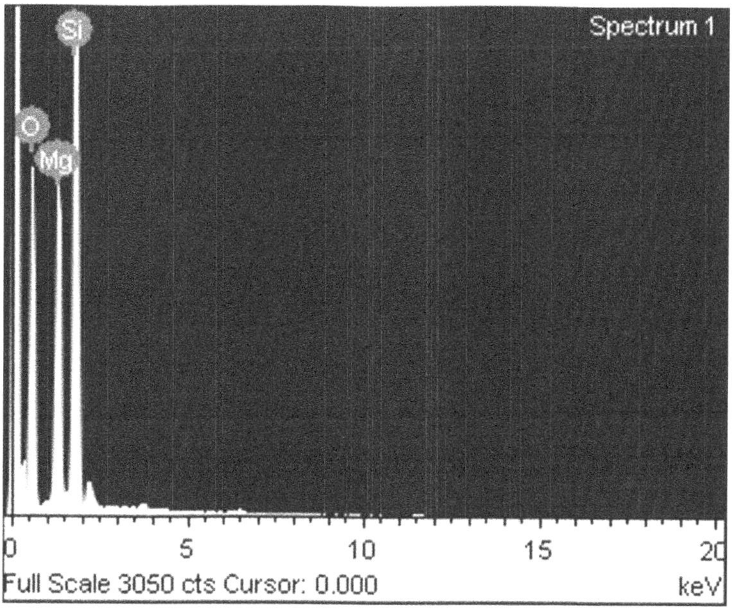

Figure 5.8. *EDS spectrum of raw SiO_2-MgO. For a color version of this figure, see www.iste.co.uk/kumar/materials.zip*

The EDS analysis shows that the weight percentages of Mg and Si contents significantly increased, whereas the O content significantly reduced due to the effect of calcination temperature (950°C), which leads to interfering of peaks. These phenomena may happen due to the marginal differences in the atomic numbers of the elements used. This same trend was also observed by calcination studies (Olivia *et al.* 2017).

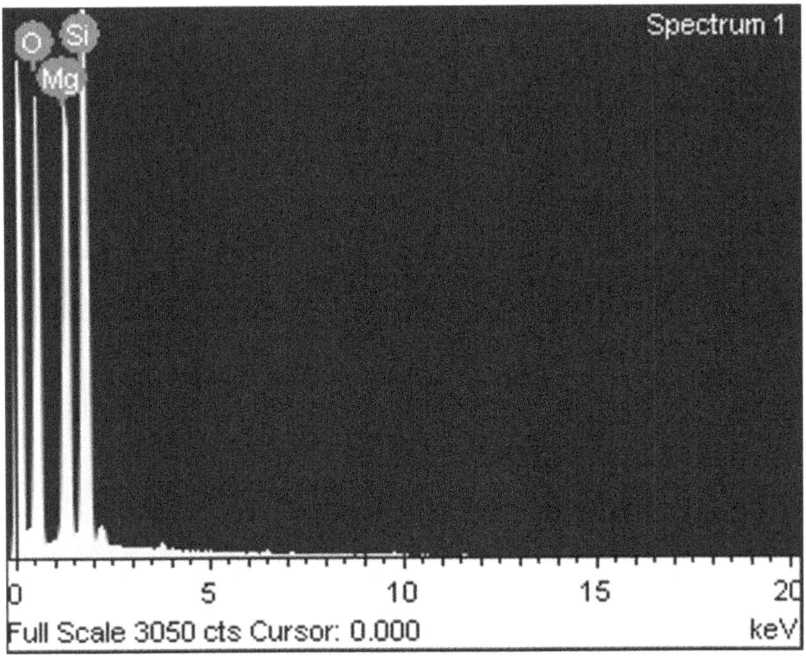

Figure 5.9. *EDS spectrum of 950°C calcinated silica-magnesium catalyst. For a color version of this figure, see www.iste.co.uk/kumar/materials.zip*

5.3.3. *Model development for catalytic cracking of polystyrene*

In this optimization study, RSM was used for the experimental factorial design, Box–Behnken design (BBD), and to evaluate the optimum conditions of process parameters. A report of the process parameters (C:F ratio, reaction temperature and reaction time) was produced by analysis of the variance (ANOVA). The ranges and their coded level of process parameters in this statistical analysis are specified in Table 5.2. The predicted responses of the liquid yields (Y) were estimated by the second-order polynomial equation expressed in equation [5.2].

Variables	Description of variables	Range and coded levels		
		Low	**Center**	**High**
		-1	**0**	**1**
A	Catalyst-to-feedstock ratio	0.25	0.5	1
B	Temperature (°C)	400	450	500
C	Reaction time (min)	30	45	60

Table 5.2. *Ranges and coded level of process parameters on liquid yield response*

The predicted responses of the liquid yield were estimated as actual factors, and mathematical models were generated from equation [5.2]. The coded equation is used to fit a second-order polynomial quadratic model using the regression analysis. The final coded expression for the liquid yield response is given in equation [5.4]:

$$\boldsymbol{Liquid\ yield\ response}\ (Y\%) = +0.0129 + 0.0012\,A + 0.0006\,B - 0.0012\,C + 0.0010\,AB + 0.0004\,AC - 0.0009\,BC - 0.0005\,A^{2} + 0.0012\,B^{2} + 0.0036\,C^{2} \quad [5.4]$$

where Y is the predicted liquid yield response, A is the C:F ratio, B is the reaction temperature and C is the time. The experimental and predicted designs of liquid yields (Y) are illustrated in Table 5.3.

Run	C:F ratio	Temperature	Time	Response of liquid yield	
				Experimental	**Predicted liquid (Y)**
1	-1	0	0	87.45	88.7
2	1	-1	0	84.32	84.43
3	-1	1	1	76.12	76.66
4	-1	1	-1	53.23	56.8
5	0	0	0	78.92	78.86
6	0	0	-1	56.89	57.7
7	-1	-1	-1	57.45	57.56
8	-1	0	-1	59.76	63.4
9	-1	1	0	75.45	74.45

10	0	1	0	67.90	68.66
11	0	0	1	72.26	73.75
12	1	0	-1	51.4	52.23
13	1	0	1	55.78	55.9
14	-1	-1	1	61.13	61.61
15	0	-1	0	69.65	72.56
16	-1	0	1	78.9	81.3
17	1	1	0	66.08	66.14

Table 5.3. *Experimental and predicted response of liquid yield (Y)*

Design-Expert 12 was used to fit the process variables, in addition to analysis of the variance (ANOVA). The ANOVA results corresponding to equation [5.4] are summarized in Table [5.4].

Source	ANOVA for the responses of liquid yield					
	Sum of squares	(df)	Mean square	F-value	p-value (Prob>F)	
Model	1,331.38	9	147.93	582.34	0.0253	Significance
A – C:F	70.15	1	70.15	971.04	0.0296	
B – Temperature	88.72	1	88.72	1230	0.02430	
C – Reaction time	36.77	1	36.77	5.089	0.0038	
AB	29.98	1	29.98	414.9	0.00128	
AC	56.48	1	56.48	78.18	0	
BC	0.7656	1	0.7656	106.06	0.02466	
A^2	213.63	1	213.63	2.96	0.005	
B^2	246.78	1	246.78	3.42	0.00153	
C^2	957.17	1	957.17	132.5	0.00024	
Residual	505.75	7	72.25	-	0	
Lack of fit	3.28	9	0.414	2.05	0	
Pure error	0.9710	1	-	-	-	
Cor total	3,541.82	16	-	-	-	

Table 5.4. *Analysis of variance (ANOVA) for the responses of liquid yield (Y)*

Table 5.4 shows the ANOVA model for the liquid yield response (Y), which is developed by equation [5.4]. The F-value of the model is 582.34, which shows the model is significant with just a 0.0253% possibility that an "F-esteem" could happen because of noise. The model p-value is less than 0.05, and the lack of fit is 2.05 (F-value), which reveals that the developed model is more significant. From an ANOVA table, the p-value is less than 0.05 for all parameters. Therefore, the developed quadratic model demonstrated that the model is more significant in terms of the F-value and p-value.

As indicated by equation [5.2], a comparative plot between the actual and predicted yield response is shown in Figure 5.10. The fit statistical model obtained from ANOVA results is summarized in Table 5.5. The adequacy precision value of 6.6392, which is greater than 4, indicates that the statistical model is a desirable one. The predicted R^2 value is 0.8702, and the adjusted R^2 value is 0.8815; these deviations showed mutual understanding among the predicted and adjusted R^2 values. Hence, the developed quadratic model can be used to explore the design space.

Parameters	**Liquid fuel (Y %)**
S.D	0.0014
ū	0.0149
C.V %	1.09
R^2	0.8606
Adjusted (R^2)	0.8815
Predicted (R^2)	0.8702
Adequacy precision	6.6392

Table 5.5. *Fit statistical model from ANOVA results for the response of liquid yield*

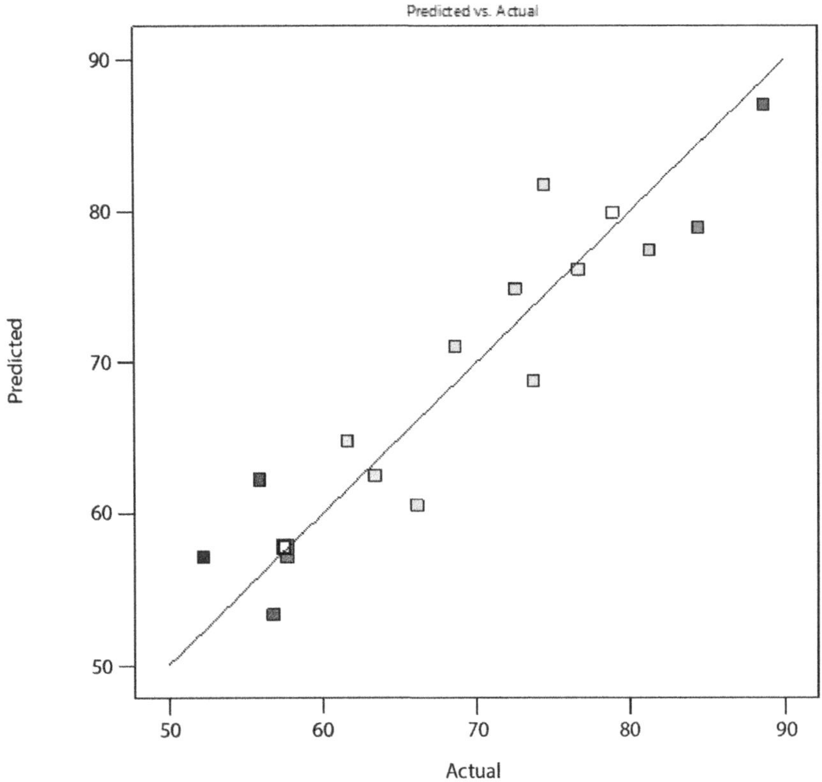

Figure 5.10. *Comparison plot between experimental and predicted data. For a color version of this figure, see www.iste.co.uk/kumar/materials.zip*

5.3.4. *Combined effect of process parameters on the response (Y)*

RSM was used to investigate the interaction effects of the operating parameters on the liquid yield response (Y) from the catalytic cracking of polystyrene.

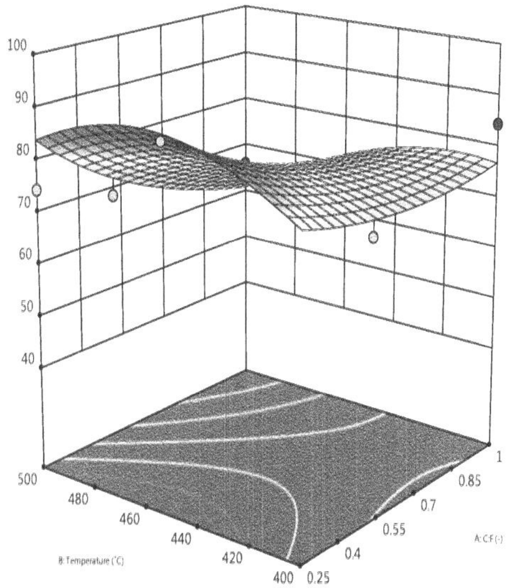

Figure 5.11. *Surface plot interaction between C:F and temperature on yield response (Y). For a color version of this figure, see www.iste.co.uk/kumar/materials.zip*

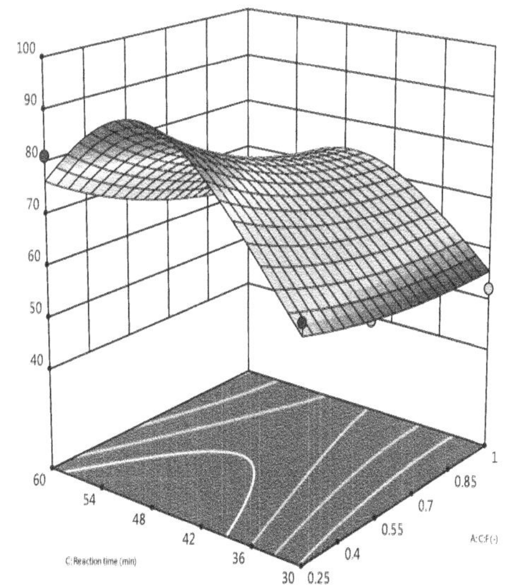

Figure 5.12. *Surface plot interaction between C:F and time on yield response (Y) . For a color version of this figure, see www.iste.co.uk/kumar/materials.zip*

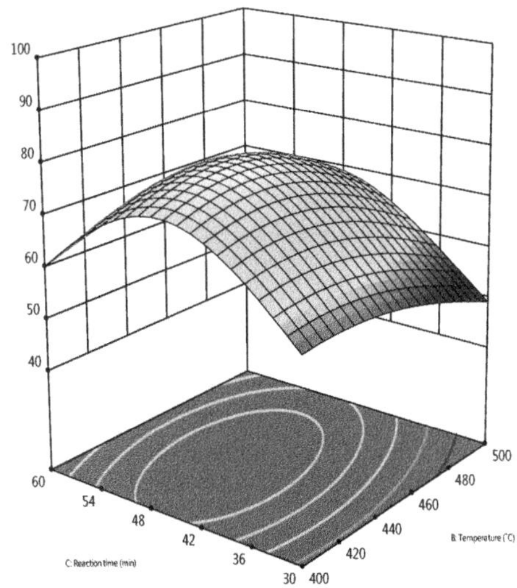

Figure 5.13. *Surface plot interaction between the temperature and the time on the yield response (Y). For a color version of this figure, see www.iste.co.uk/kumar/materials.zip*

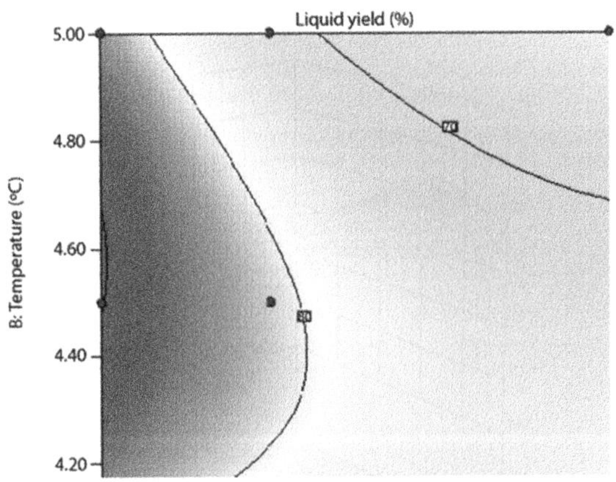

Figure 5.14. *Contour plot interaction between the C:F ratio and the temperature on the yield response (Y). For a color version of this figure, see www.iste.co.uk/kumar/materials.zip*

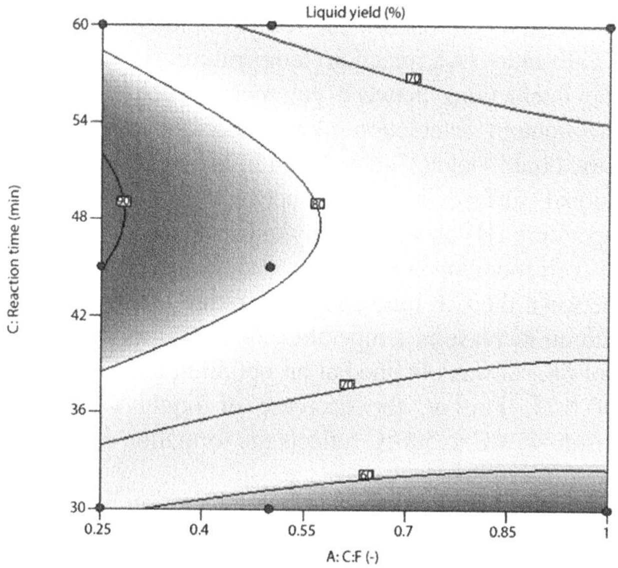

Figure 5.15. *Contour plot interaction between the C:F ratio and time on the yield response (Y). For a color version of this figure, see www.iste.co.uk/kumar/materials.zip*

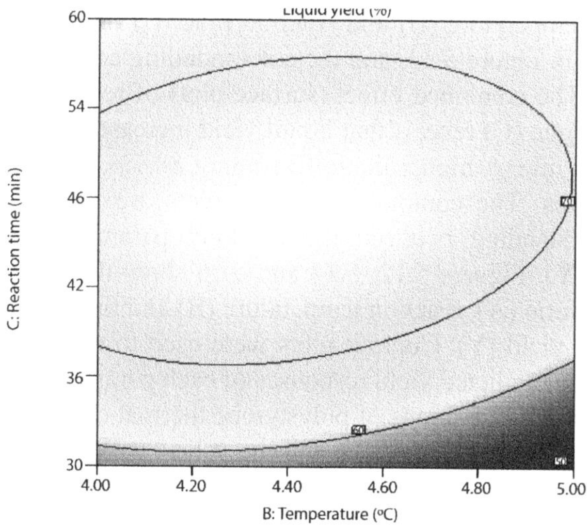

Figure 5.16. *Contour plot interaction between the temperature and time on the yield response (Y). For a color version of this figure, see www.iste.co.uk/kumar/materials.zip*

Based on the ANOVA results, the results of interaction between process parameters, C:F ratio (A), reaction temperature (B) and time (C), were obtained. The interactions between parameters such as surface interaction plots and the contour interaction plots under the optimum condition for maximizing the liquid yield (Y) are shown in Figures 5.11–5.16. The effect of three-dimensional surface interaction between the C:F ratio (A) and the reaction temperature (B) on the liquid yield response (Y) is shown in Figure 5.11, and the corresponding contour plot is shown in Figure 5.14. The surface interaction between the C:F ratio and temperature shows that the liquid yield increased with an increase in temperature from 400 to 450°C. Also, a higher liquid yield of 88.7% was obtained at an optimum temperature of 450°C and C:F ratio of 0.25. Further, the decrease of liquid yield with increasing temperature from 450°C to 500°C reflects the formation of char residues.

The effect of three-dimensional surface interaction between the C:F ratio (A) and the reaction time (C) on the liquid yield response (Y) is shown in Figure 5.12, and the corresponding contour plot is shown in Figure 5.15. The surface interaction between C:F ratio and time (C) indicates that the liquid yield increased with decreasing C:F ratio from 1 to 0.25 and that a higher liquid fraction of 88.7% was obtained at the optimum time of 45 minutes and 0.25 C:F ratio. The effect of three-dimensional surface interaction between the reaction temperature (B) and reaction time (C) on liquid yield response (Y) is shown in Figure 5.13, and the corresponding contour plot is shown in Figure 5.16. The combined effect (surface plot) of reaction temperature (B) and reaction time (C) reveals that liquid yield increased with increasing time from 30 to 45 min. A higher liquid fraction of 88.7% was obtained at 450°C and 45 minutes. The contour interaction plots are used for facilitating a mutual understanding between the cracking parameters for maximizing liquid yield (Y). Figures 5.12, 5.14 and 5.16 show the contour plots effect between C:F ratio (A), reaction temperature (B) and time (C) on the response of liquid fuel yield (Y). Contour plots were used to relate the experimental values with the predicted yield responses of each parameter through equation [5.4]. The optimal conditions of polystyrene thermal cracking process intend to achieve maximum liquid yield, and the values were found to be a 50 min reaction time, 425°C temperature, 500 g weight and 4 mm plastic size (Selvaganapathy *et al.* 2020). The obtained results were better satisfied with previous research on catalytic cracking of high-density polyethylene, which reveals that the optimum conditions are 450°C temperature, 0.341 catalyst acidity and 1:4 C:F ratio, which produced 78.7% liquid yield (Kumar and Singh 2014).

5.3.5. *Characterization of liquid yield*

a) *FTIR analysis of liquid yield*

FTIR analysis of liquid yield is a significant characterization technique for detecting the presence of functional groups. Upon interaction with infrared light, the chemical structure of the liquid sample will bend, producing sharp and stretched radiations in a wavelength range of 500 cm^{-1} to 4,000 cm^{-1}. The FTIR spectra of the liquid yield derived from silica-magnesium cracking of polystyrene wastes are shown in Figure 5.17. The alkene group is analyzed at a wavenumber of 3,080 cm^{-1} with =C–H stretching vibrations. The alkane groups are analyzed at 2,965 cm^{-1} and 2,862 cm^{-1} with C-H stretching vibrations. The C=O stretching at a wavenumber of 1,644 cm^{-1} reveals the presence of aldehydes. The aromatic compounds are analyzed at 1,575 cm^{-1} with C=C stretching. The CH_3C–H bending vibrations at 1,357 cm^{-1} show the presence of alkane groups. The alkyl halide is analyzed at 1,380 cm^{-1}, 1,257 cm^{-1} and 1,175 cm^{-1} with C–F stretching vibration. =C–H bend stretching analyzed at 998 cm^{-1}, 983 cm^{-1} and 821 cm^{-1} shows the presence of alkenes. The alkyl halide is analyzed at 685 cm^{-1} and 584 cm^{-1} with C–Br stretching vibrations. The FTIR spectra of the liquid yield derived from silica-magnesium cracking of polystyrene wastes show that the hydrocarbon consists of aliphatic, aromatic and olefinic compounds with major constituents of aromatic groups. The same results were reported in our previous report (Selvaganapathy *et al.* 2020).

b) *GC/MS analysis of liquid yield*

The hydrocarbon compound present in the liquid sample was characterized using GC/MS analysis, as shown in Figure 5.18. In accordance with retention time, highest peak areas of chromatograms appeared in the liquid yield derived from silica-magnesium cracking of polystyrene, such as the retention time of 2.699 s for cyclopropane, 4.152 s for 1,2-dimethyl, 4.461 s for n-hexane, 5.125 s for ethyl acetate, 7.321 s for 1-heptene, 10.701 s for 2,4-dimethyl-1-heptene, 11.529 s for 1-nonene, 11.667 s for 1-nonane, 13.068 s for 1-decene, 14.422 s for 1-undecene, 15.843 s for 1-dodecene, 17.624 s for 1-tridecene and 18.091 s for octane, 4-methyl and more hydrocarbon compounds with the smallest peak. From the GC/MS analysis, it was observed that the liquid sample contains about 93 compounds and mostly contained gasoline-range products within C_3–C_{13}. Similar results were also reported by several works (Kumar and Singh 2014; Selvaganapathy *et al.* 2020).

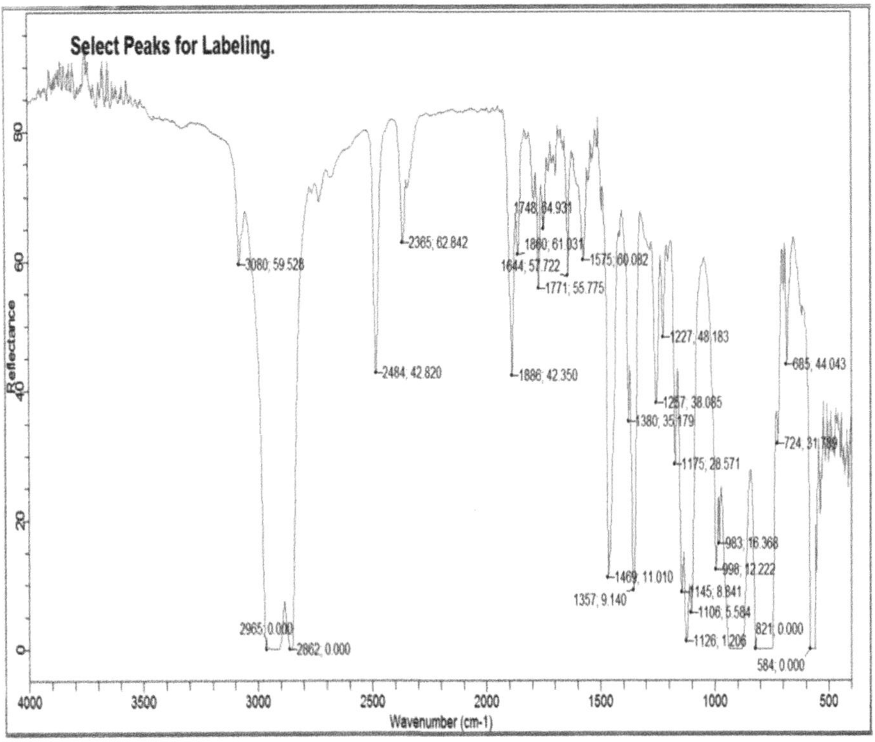

Figure 5.17. *FTIR analysis of silica-magnesium cracking of polystyrene. For a color version of this figure, see www.iste.co.uk/kumar/materials.zip*

c) *Physico-chemical properties of liquid fuel*

The physico-chemical analyses of the liquid sample derived from silica-magnesium cracking of polystyrene were conducted, and the results were compared with the standard diesel properties. The comparison between the obtained liquid sample and the standard diesel properties is illustrated in Table 5.6.

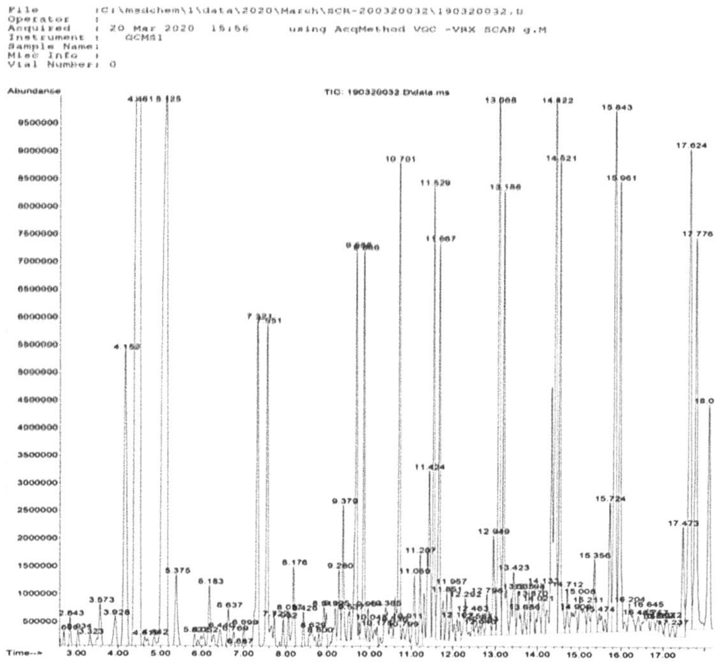

Figure 5.18. *GC-MS analysis of silica-magnesium cracking of polystyrene. For a color version of this figure, see www.iste.co.uk/kumar/materials.zip*

Property	Silica-magnesium cracking of polystyrene	Polystyrene liquid fuel (Selvaganapathy *et al.* 2020)	Diesel (Sharuddin *et al.* 2016)
Density (g/cc)	0.822	0.792	0.807
Kinematic viscosity (CSt)	1.0	1.78	1.9-4.1
Pour point (°C)	+5	+2	+6
Flash point (°C)	35	32	52
Fire point (°C)	42	49	-
GCV (MJ/kg)	44.45	43.65	43.0

Table 5.6. *Physico-chemical analysis of liquid sample with standard diesel property*

The density of the derived liquid was found to be 0.822 g/cc, which is almost equivalent to the commercial diesel density. The observed density value is higher than the polystyrene liquid product. The kinematic viscosity of the liquid sample was 1.0 CSt, extremely lower than that of the regular diesel (Selvaganapathy *et al.* 2020). The pour point is considered to be relatively larger than the polystyrene liquid yield and equivalent to the diesel-range products. Pour point of the fuel less than -15°C leads to freezing troubles in cold regions (Kumar and Singh 2014). The flashpoint was observed at 35°C, which is similar to polystyrene liquid and lower than the commercial diesel. The GCV of the liquid yield was estimated to be 44.45 MJ/kg, which is similar to the commercial diesel (Sharuddin *et al.* 2016).

Hence, the pyrolysis liquid fuel was most suitable for transportation fuel instead of commercial gasoline and diesel fuel even in cold regions.

5.3.6. *Factors affecting catalytic cracking of polystyrene*

a) *Effect of C:F ratio on silica-magnesium cracking of polystyrene*

In order to optimize the experimental parameters on catalytic cracking of polystyrene, different ratios of the catalysts and polystyrene were introduced in a bench-scale pyrolytic reactor at 400–500°C for 45–90 min. Different ratios of silica-magnesium catalyst and polystyrenes such as 1:4, 1:2, 3:4 and 1:1 were investigated to optimizing the catalyst-to-feedstock ratio. The effect of the C:F ratio on the catalytic cracking product distribution is shown in Figure 5.19.

Each ratio of silica-magnesium catalyst to polystyrene was investigated, and it was found that a higher liquid yield of 91.2% was observed at the ratio of 1:4. A compatible observation has also been reported on the catalyst-to-feedstock ratio, and it was reported that the kaolin catalyst was the most efficient catalyst for the conversion of polypropylene into gasoline-, diesel- and kerosene-range products. The maximum liquid yield obtained was 87.5% at the 1:3 ratio of the catalyst with plastic (Panda *et al.* 2013).

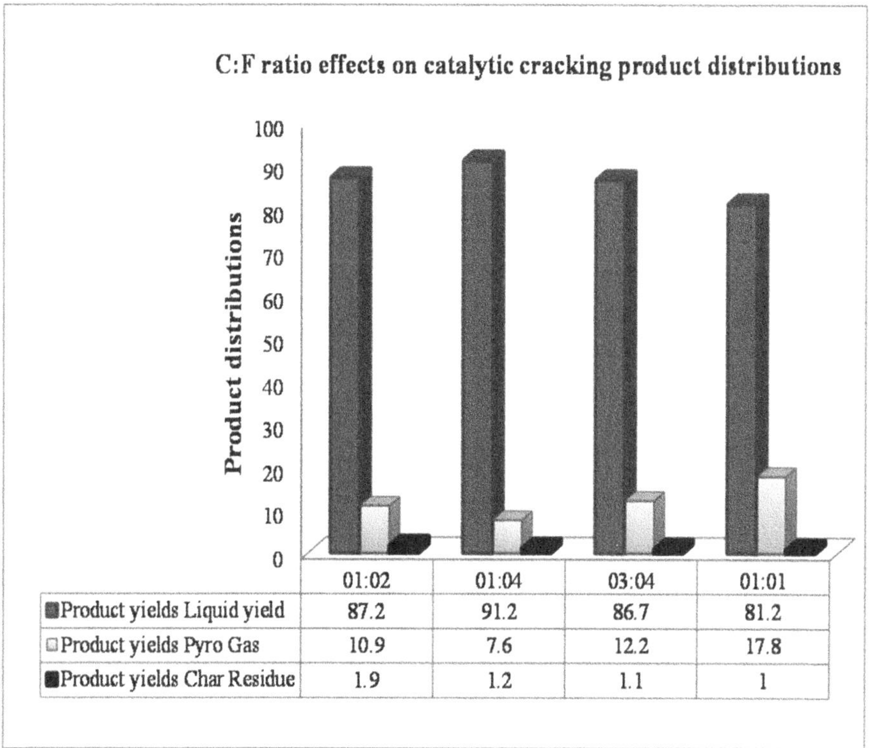

	01:02	01:04	03:04	01:01
Product yields Liquid yield	87.2	91.2	86.7	81.2
Product yields Pyro Gas	10.9	7.6	12.2	17.8
Product yields Char Residue	1.9	1.2	1.1	1

Figure 5.19. *Effect of C:F ratio on silica-magnesium cracking of polystyrene. For a color version of this figure, see www.iste.co.uk/kumar/materials.zip*

b) *Effect of temperature on silica-magnesium cracking of polystyrene*

To investigate the optimum reaction temperature for converting the maximum quantity of liquid fuel, the waste polystyrene was catalytically degraded at 400, 425, 450, 475 and 500°C. The effect of temperature on the catalytic cracking process is presented in Figure 5.20. The figure demonstrates that the liquid yield increased with an increase in reaction temperature from 400 to 450°C. Further, the liquid fuel yield decreased with increasing reaction temperature from 450 to 500°C. At the optimum temperature of 450°C, the conversion of polystyrene to liquid fuel was 91.2%, while the conversion into a pyro-gas product was 7.6%. The results obtained fit well with the results of other research groups (Kumar and Singh 2011; Panda *et al.* 2013; Nisar *et al.* 2019). J. Nisar *et al* . (2019) reported the effect of temperature on product distribution of polystyrene pyrolysis in an indigenous pyrolyzer within the temperature range of 360–420°C. It was found that an increase in liquid yield was detected with increasing temperature from 360 to 420°C and an increase in gaseous products was detected with increasing temperature range above 420°C (Nisar *et al.* 2019).

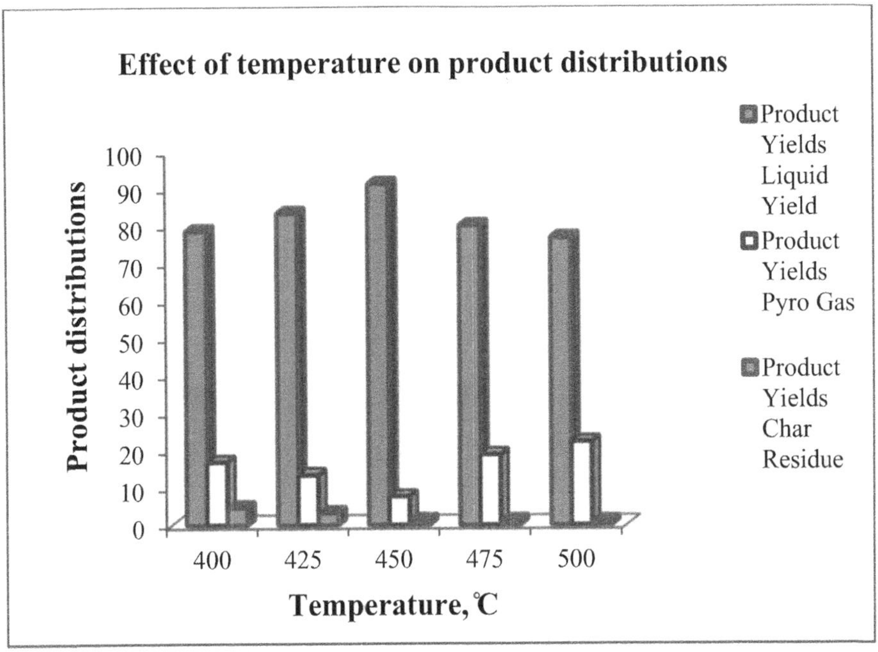

Figure 5.20. *Effect of the temperature on silica-magnesium cracking of polystyrene. For a color version of this figure, see www.iste.co.uk/kumar/materials.zip*

c) *Effect of time on silica-magnesium cracking of polystyrene*

The effect of reaction time on silica-magnesium cracking of polystyrene was also studied at 450°C and 1:4 C:F ratio, and the observed results are shown in Figure 5.21. The figure shows the maximum conversion of the liquid yield of about 91.2% at an optimum time of 45 min. Further, the liquid fuel yield increases with increasing time from 0 to 45 min and decreases with increasing the time from 45 to 90 min. The results show that a further increase in reaction time causes the formation of a more gaseous fraction. The optimum reaction temperature in our investigations is in good agreement with the reported literature, and it stated that both lower reaction time and very long reaction time are most preferable for gaseous and char residue products, whereas the intermediate reaction time is most suitable for

liquid products (Kumar and Singh 2011; Panda *et al.* 2013; Nisar *et al.* 2019).

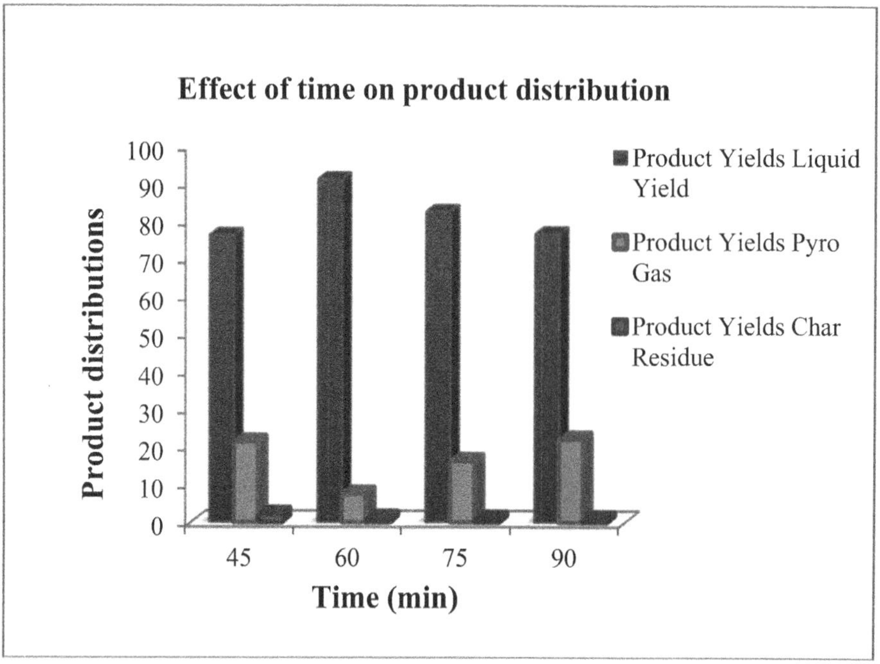

Figure 5.21. *Effect of time on silica-magnesium cracking of polystyrene. For a color version of this figure, see www.iste.co.uk/kumar/materials.zip*

5.4. Conclusion

The optimization study on the catalytic cracking of plastic waste was successfully demonstrated by the response surface method using the Box–Behnken design (BBD). The optimum process conditions of the catalytic cracking process are 0.25 C:F ratio, 450°C operating temperature and 45 min time, which produced a higher fraction of liquid yield of 88.7%. The predicted and adjusted R^2 values are 0.8702 and 0.8815 respectively; these deviations showed a mutual relationship between the predicted and adjusted R^2 values. Hence, the developed quadratic model can be used to explore the design space.

Moreover, an experimental investigation on silica-magnesium cracking of polystyrene was conducted. Previously, the raw SiO_2-MgO material was

calcinated (950°C) and characterized using SEM-EDS analysis. The optimum experimental conditions for the polystyrene catalytic cracking process are 1:4 C:F ratio, 450°C reaction temperature and 45 min reaction time, which yielded a higher liquid fraction of 91.2%. These similarities provided a reasonable agreement between experimental and statistical operating conditions.

The GC/MS analysis of the liquid sample shows that silica-magnesium cracking of polystyrene is used for the production of gasoline-range products within C_3–C_{13}. Also, the FTIR spectra show that the produced hydrocarbon mainly consists of aromatic compounds. Hence, it could be concluded that the developed quadratic model well fit the experimental domains and that the predicted operating conditions are most suitable for conducting catalytic cracking experiments.

5.5. References

Alla, M.M.G. and Ali, S.O.A. (2014). Simulation and design for process to convert plastic waste to liquid fuel using Aspen Hysys program. *International Journal of Engineering Research & Technology*, 1(6), 270–274.

Amoodi, N.A., Kannan, P., Shoabi, A.A., and Srinivasakannan, C. (2013). Aspen plus simulation of polyethylene gasification under equilibrium conditions. *Chemical Engineering Communications*, 200, 977–992.

Awasthi, A.K., Shivashankar, M., and Majumder, S. (2017). Plastic solid waste utilization technologies: A review. *IOP Conference Series: Materials Science and Engineering*, 263, 1–13.

Babajide, O., Petrik, L., Musyoka, N., Amigun, B., and Ameer, F. (2010). Use of coal fly ash as a catalyst in the production of biodiesel. *Petroleum & Coal*, 52(4), 261–272.

Cornell, J.A. (1990). *How to Apply Response Surface Methodology*. American Society for Quality Control, Wisconsin.

Istadi, I., Suherman, S., and Buchori, L. (2010). Optimization of reactor temperature and catalyst weight for plastic cracking to fuels using response surface methodology. *Bulletin of Chemical Reaction Engineering & Catalysis*, 5, 103–111.

Jouhara, H., Ahmad, D., van den Boogaert, I., Katsou, E., Simons, S., and Spencer, N. (2018). Pyrolysis of domestic based feedstock at temperatures up to 300°C. *Thermal Science and Engineering Progress*, 5, 117–143.

Krishankant, J.T., Bector, M., Rajesh Kumar, C., and Taneja, J. (2012). Application of response surface modeling for determination of flux consumption in submerged arc welding by the effect of various welding. *International Journal of Mechanical Engineering and Robotics Research*, 1, 248–256.

Kumar, S. and Singh, R.K. (2011). Recovery of hydrocarbon liquid from waste high density polyethylene by thermal pyrolysis. *Brazilian Journal of Chemical Engineering*, 28(4), 659–667.

Kumar, S. and Singh, R.K. (2014). Optimization of process parameters by response surface methodology (RSM) for catalytic pyrolysis of waste high-density polyethylene to liquid fuel. *Journal of Environmental Chemical Engineering*, 2, 115–122.

Kunwar, B., Cheng, H.N., Chandrashekaran, S.R., and Sharma, B.K. (2016). Plastics to fuel: A review. *Renewable and Sustainable Energy Reviews*, 54, 421–428.

Mahalik, K., Sahu, J.N., Patwardhan, A.V., and Meikap, B.C. (2010). Statistical modelling and optimization of hydrolysis of urea to generate ammonia for flue gas conditioning. *Journal of Hazardous Materials*, 182, 603–610.

Miandad, R., Barakat, M.A., Aburiazaiza, A.S., and Rehan, M. (2016). Catalytic pyrolysis of plastic waste: A review. *Process Safety and Environmental Protection*, 102, 822–838.

Ministry of Housing and Urban Affairs (2019). Annual report [Online]. Available at: http://mohua.gov.in/.

Miranda, M., Pinto, F., Gulyurtlu, I., Cabrita, I., Nogueira, C.A., and Matos, A. (2010). Response surface methodology optimization applied to rubber tire and plastic wastes thermal conversion. *Fuel*, 89, 2217–2229.

Montgomery, D.C. (2012). *Design and Analysis of Experiments*, 8th edition. John Wiley and Sons, New York.

Nisar, J., Ali, G., Shah, A., Iqbal, M., Khan, R.A., Sirajuddin, Anwar, F., Ullah, R., and Akhter, M.S. (2019). Fuel productions from waste polystyrene via pyrolysis: Kinetics and products distribution. *Waste Management*, 88, 236–247.

Olivia, R., Jamarun, N., Arif, S., and Sirin, Y.A. (2017). The utilization of dolomite as catalyst in biodiesel production. *Rasāyan Journal of Chemistry*, 10(1), 160–164.

Panda, A.K. and Singh, R.K. (2013). Experimental optimization of process for the thermo-catalytic degradation of waste polypropylene to liquid fuel. *Advances in Energy Engineering (AEE)*, 1(3), 74–84.

Patni, N., Shah, P., Agarwal, S., and Singhal, P. (2013). Alternative strategy for conversion of waste plastic to fuels. *ISRN Renewable Energy*, 2013, 1–7.

Pinto, F., Paradela, F., Gulyurtlu, I., and Ramos, A.M. (2013). Prediction of liquid yields from the pyrolysis of waste mixtures using response surface methodology. *Fuel Processing Technology*, 116, 271–283.

Selvaganapathy, T., Muthuvelayudham, R., and Jayakumar, M. (2020). Process parameter optimization study on thermolytic polystyrene liquid fuel using response surface methodology (RSM). *Materials Today: Proceedings*, 26, 2729–2739.

Shah, J., Jan, M.R., Mabood, F., and Jabeen, F. (2020). Catalytic pyrolysis of LDPE leads to valuable resource recovery and reduction of waste problems. *Energy Conservation and Management*, 51, 2791–2801.

Sharuddin, S.D.A., Abnisa, F., and Wan Daud, W.M.A. (2016). A review on pyrolysis of plastic wastes. *Energy Conversion and Management*, 115, 308–326.

Vas, J.P., Ramya, M., Dsilva, A.L., Serrao, E., Demash, F., and Dsa, J.D. (2017). Production of high grade liquid fuel for CI engine by thermo-catalytic cracking of waste plastic. *Energy and Power*, 7(3), 81–87.

PART 3

Analysis

6

FEA Comparative Studies on Heat Flux and Thermal Stress Analysis during Conduction Mode and Keyhole Mode in the Laser Beam Welding

Laser welding is one of the key weld processes used for joining metals in which high joint strength is obtained with low heat-affected zone. This chapter presents suitable models used for the application of a Gaussian beam profile and the frustum model to conduction mode welding and keyhole laser welds. Temperature and stress analysis were carried out within and around the weld region. This chapter discusses the analytical comparative approximation of different model approaches applicable to the laser weld process. The parametric study and analysis can effectively be used in order to design and optimize the weld process parameters for engineers widely employed in the manufacturing sector like nuclear, aeronautical and chemical plants.

6.1. Introduction

Laser welding is one of the advanced joining processes used for steels in various applications such as nuclear reactors, chemical pipelines, automobile and aeronautical sectors. The advantages of low heat input, high-speed production rates and narrow heat-affected zone formation are considered for typical applications. For fabrication, thin sections are joined using low power lasers (conduction mode), while thick sections are joined using high power lasers (keyhole mode) (Nath *et al.* 2002). In the conduction mode welding

---

Chapter written by Harinadh VEMANABOINA, Suresh AKELLA and Ramesh Kumar BUDDU.

process, which employs low laser input power (<1 kW and lower heat input), weld bead to depth ratios are low (<1) and the welding zone is relatively wide compared to the depth which forms a lower heat affected zone. In the keyhole mode, the molten pool flows with high momentum due to the plasma keyhole and vapor phase formation, and experiences narrow and deep penetration of the weld bead (Akella *et al.* 2016). The temperature and stress analysis in both cases is useful for predicting and examining the structural integrity of components (Vemanaboina *et al.* 2014). In addition, the process includes metallurgical effects with phase transformation with the associated melting and cooling processes of the used material. During the laser weld process, monitoring of weld pool temperature and further stress analysis measurements pose a challenge to measurement techniques due to the associated complex instrumentation requirements. (Zain-ul-abdein *et al.* 2010; Akella *et al.* 2014, 2016). However, simulation of the temperature and stress analysis is conducted using mathematical and numerical material modeling techniques (Ki *et al.* 2002; Akella *et al.* 2014). However, each approach has a limitation due to the assumptions of conduction and convection made in the formulated thermal models (Kaplan 1994; Zeng-Rong *et al.* 2007; Fabbro 2010). Experimental analysis was carried out using the CO_2 laser welding process for both the conduction and keyhole modes for SS304 materials by measuring the laser power coupling efficiency, which is crucial for welding (Nath *et al.* 2002; Vemanaboina *et al.* 2018a, 2019a).

In the laser welding process for thin or thick sections, it is necessary to understand the state of the structure exposed to thermal gradients, which are responsible for the presence of residual stress that causes weld deformation (Belhadj *et al.* 2010). This chapter focuses on the simulation of the conduction and keyhole modes using the Gaussian and frustum models based on finite element analysis. The transient thermal and stress analysis were carried out for each case using Gaussian and frustum models and then compared.

6.2. Heat in laser welding

As shown in Figure 6.1, a welding process joins two parts by applying heat at the surface intersection. Heat is conducted between the two parts laterally. The immediate neighboring parts heat faster and reach the melting phase before the weld heat moves ahead. During the welding process, melt pool flow is influenced by parameters such as weld speed and heat input, which governs the weld bead, heat affected zone and weld zone formation, and its dimensions, which are prominent features in Laser and TIG welding processes. However, beam welding processes such as laser and electron beam welding contain very high energy density that is focused at the center (Rosenthal 1946; Anupamadev *et al.* 2020). Here, new physical activities evolve: for example, the molten metal is no longer static and flows under the influence of gravity and the opposing surface tension; of more importance is the vaporization of the molten metal and the formation of the plasma state of the vaporized metal. Along with the shielding gas, the plume holds the laser beams within the cavity that expands, and forms the keyhole due to the thermo-capillary effect. As modeled by Kaplan (1994; Goldak and Akhlagi), due to multiple reflections and absorptions, the retained heat of laser beams is transferred by conduction to the neighboring base metals laterally, which helps in accelerated melting and metal joining operation, and a portion of heat is released back to the surface. In this work, the laser heat on the surface is divided into parts: one part acts longitudinally on the surface, and the other part is the keyhole heat that acts laterally on the keyhole. The keyhole heat can be formulated as conical, as a frustum or Gaussian. In this study, it is modeled as cylindrical, so the total heat is a combination of Gaussian on the surface and cylindrical at the keyhole.

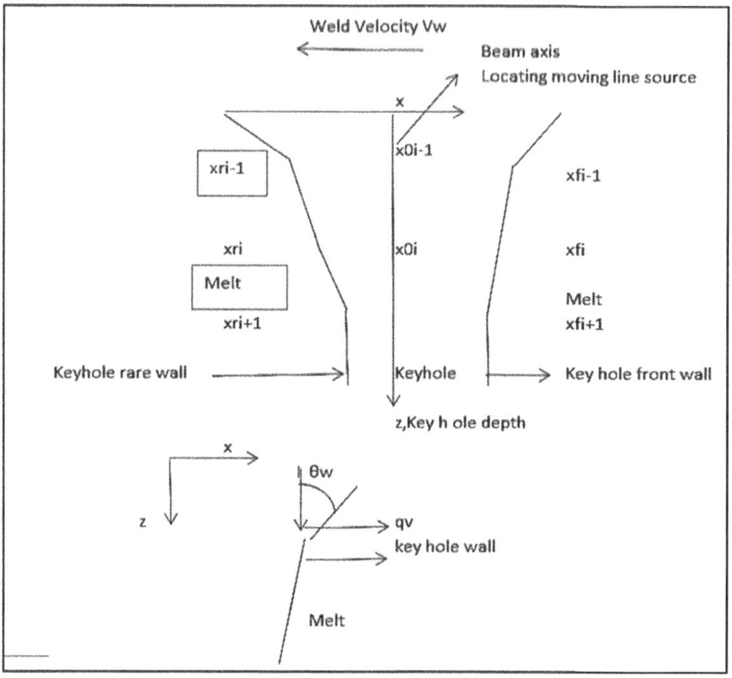

Figure 6.1. *Representation of Kaplan's keyhole model*

6.3. Modeling

Gaussian beam and frustum profile models were used for the transient thermal analysis. Thermal loads were applied on the surface of the bead area. A single-pass butt welding process was carried out using APDL coding with a heat flux of 1,200 W (Vemanaboina *et al.* 2014). The welding simulation time was increased up to 1,000 sec, in order to allow the weldment to cool to the ambient temperature. Solid 70 element was used for this simulation; the element supports both transient thermal and static structural analysis (Vemanaboina *et al.* 2018b, 2019b). The model dimensions used for the present simulation are presented in Figure 6.2. The bead width of 2 mm was considered for the analysis. The mesh size of 0.003 mm was used, as shown in Figure 6.3.

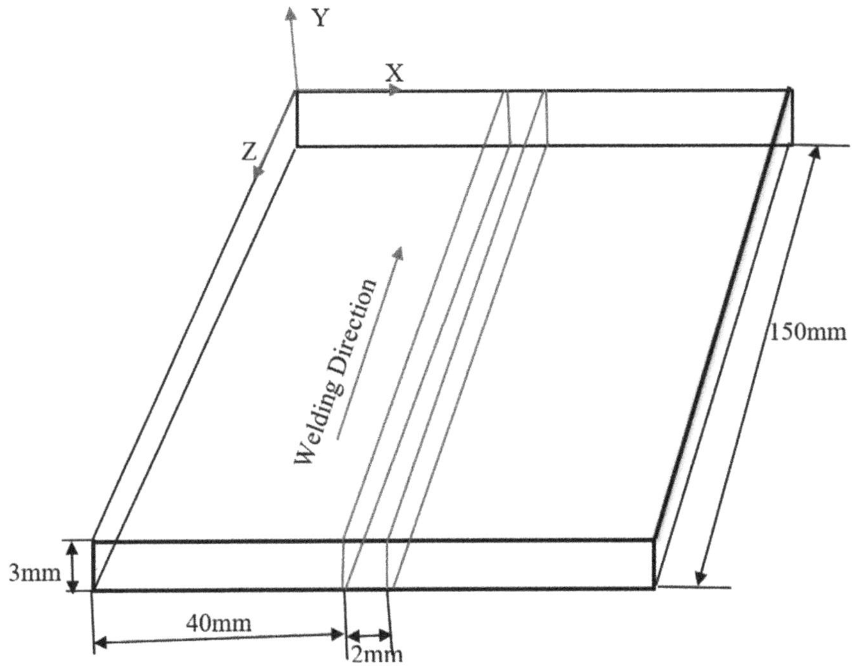

Figure 6.2. *Geometry of the model. For a color version of this figure, see www.iste.co.uk/kumar/materials.zip*

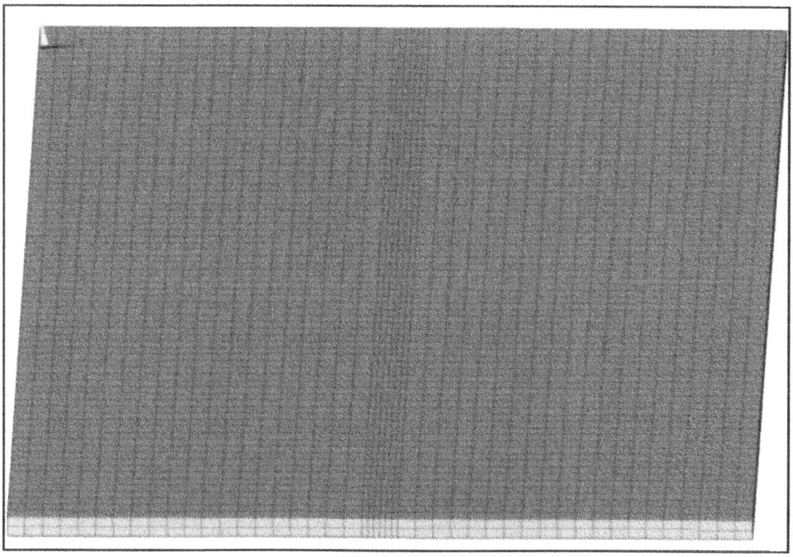

Figure 6.3. *Meshing of the model. For a color version of this figure, see www.iste.co.uk/kumar/materials.zip*

The convective heat transfer coefficient, h (W/m^2K), is given for surfaces where the plate is exposed to the ambient temperature. However, from the literature, the radiation boundary condition may be added to obtain a modified convective heat transfer coefficient (h_r). Here, ε is the emissivity of surface radiating and σ is the Stefan Boltzmann constant (5.67×10^{-8} W/m^2K^4). As the radiation depends on a fourth power difference in temperature to satisfy the radiation boundary condition, it is combined with convective heat flux using a modified coefficient, h_r, for hot rolled steel plates with an error of about 5%. Radiation inclusion will increase the solution time by about three times and hence combined with convection:

$$H = 24.1\times10^{-4}\ \varepsilon T^{1.61} \quad [6.1]$$

6.4. Results and discussion

6.4.1. *Keyhole model*

The surface heat flux area is used for the thermal analysis; the bead width is wider for the arc welding process and narrow for laser welds. To evaluate the effect of the base width variation, the Gaussian distribution for 1,200 W is studied. The different Gaussian shapes are shown in Figure 6.4.

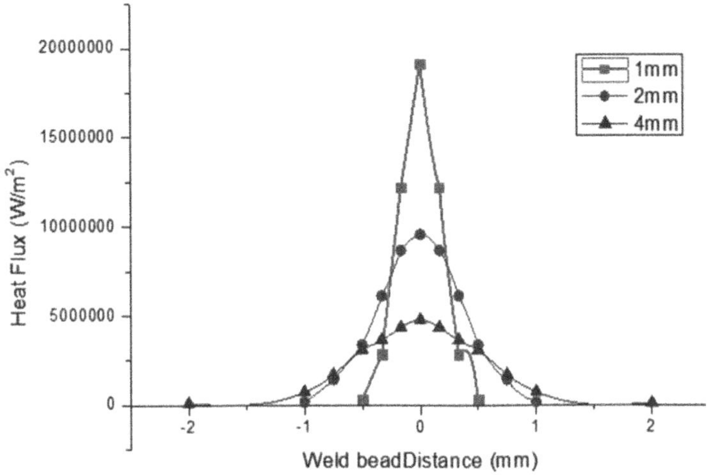

Figure 6.4. *Gaussian power distribution. For a color version of this figure, see www.iste.co.uk/kumar/materials.zip*

Then, the model was tested for full plate welding, which joins symmetric plates subjected to symmetric boundary conditions of clamping, convection from plate surfaces as well as heat flux induction. This symmetry allows the modeling of a half-plate to represent the welding process in terms of output parameters such as temperature, stresses and distortion distributions. The temperature distribution of a 1,200 W heat flux model both as a full plate and a half-plate is shown in Figure 6.5, where the temperatures of the full model and the half-plate show similar results.

The advantage of a symmetric model is to obtain half the model meshing and hence faster evaluation, as well as to improve computer space and speed. However, in the keyhole modeling, this also becomes a requirement. The keyhole heat input is on the thickness which is lateral to the two plates joining face. The advantage of keyhole formation in laser beam welding is also shown in Figure 6.5. The temperature when 1,200 W heat is given on surface is about 1,080°C compared to about 1,520°C. This result is important to understand the difference in surface heat in TIG welding compared to laser beam welding with keyhole formation. After cooling down to ambient temperature, residual stress is estimated, as shown in Figure 6.6.

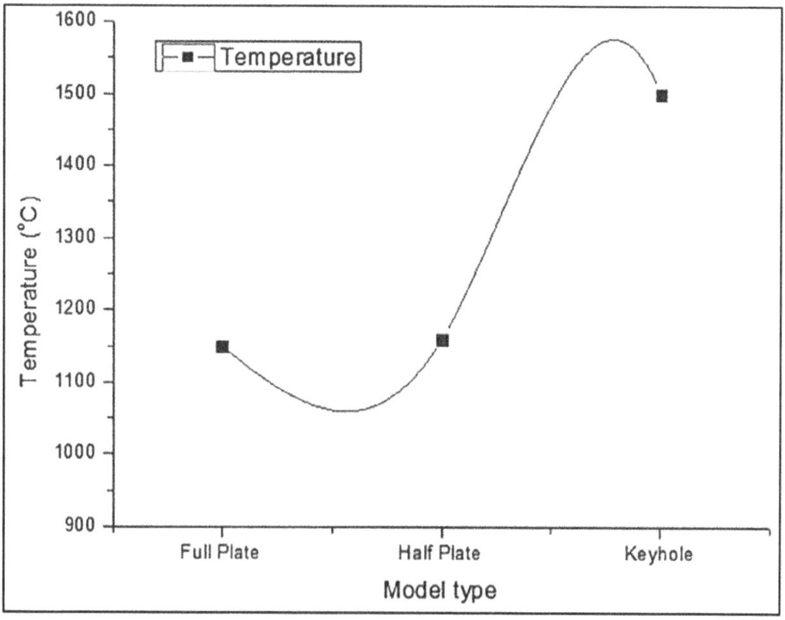

Figure 6.5. *Temperature distributions for various models. For a color versionof this figure, see www.iste.co.uk/kumar/materials.zip*

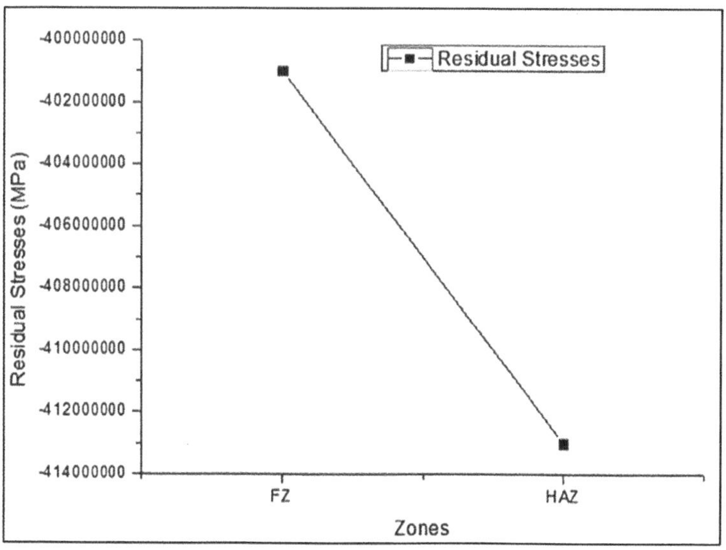

Figure 6.6. *Comparison of residual stresses in the keyhole model. For a color version of this figure, see www.iste.co.uk/kumar/materials.zip*

Residual stress is far below (280 MPa) the yield stress of the SS304 material. However, as expected, residual stress in the fusion zone is the highest and higher than the heat-affected zone. This is also expected as the temperature is the highest in the fusion zone and higher than the heat-affected zone, as shown in Figure 6.7.

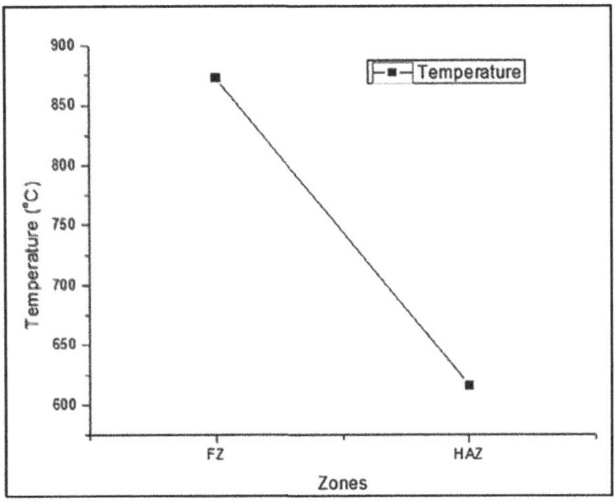

Figure 6.7. *Comparison of temperatures in the keyhole model. For a color version of this figure, see www.iste.co.uk/kumar/materials.zip*

Another model comparison made was to compare heat fluxes as a frustum or Gaussian. In Figure 6.8, we can see that the Gaussian distribution gives better heat to the weld zone compared to the frustum distribution. The temperature difference of 200°C was observed in Gaussian heat flux. This comparison was made for three measurements of widths: 1 mm, 2 mm and 4 mm. However, as expected, the higher temperature would lead to higher residual stress in the case of a Gaussian distribution than that for a frustum distribution, for the same heat of 1,200 W, as shown in Figure 6.9.

6.4.2. *Conduction model*

6.4.2.1. *Temperature distribution*

Conduction modeling was carried out on various bead widths (4 mm, 2 mm and 1 mm) with various power inputs (100 W, 400 W, 800 W and 1,200 W). All of the heat fluxes were modeled as a Gaussian distribution, as shown in Figure 6.10. This power variation was made for welding small thickness parts (less than 1 mm), which requires proportionally less power, as shown in Figure 6.11.

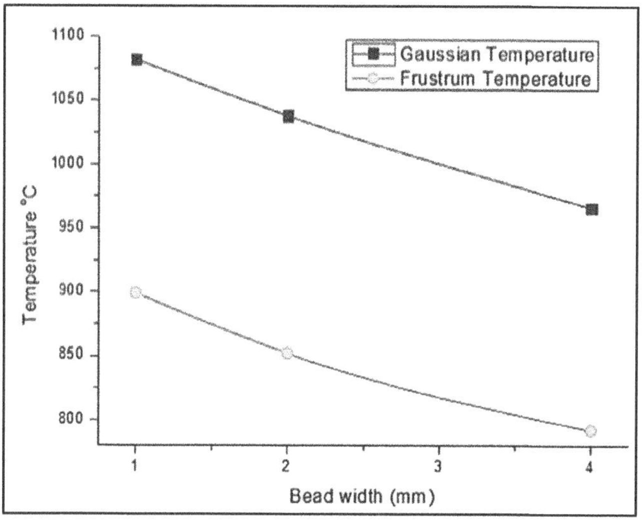

Figure 6.8. *Thermal results for Gaussian and frustum heat flux. For a color version of this figure, see www.iste.co.uk/kumar/materials.zip*

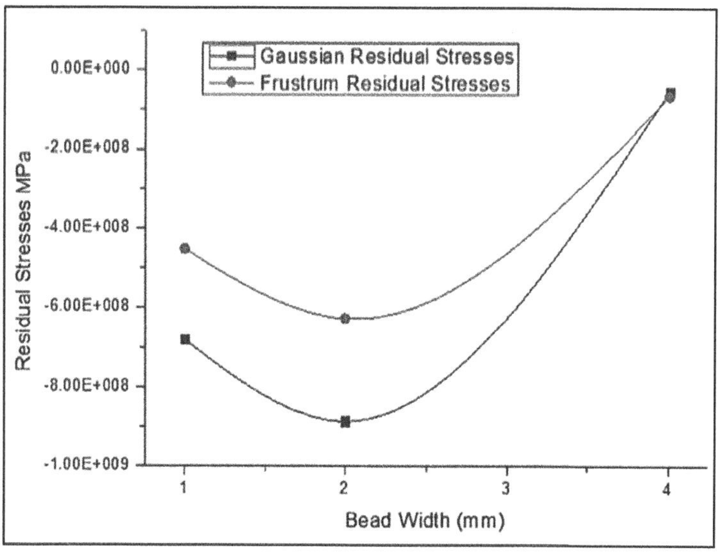

Figure 6.9. *Stress results for Gaussian and frustum heat flux. For a color version of this figure, see www.iste.co.uk/kumar/materials.zip*

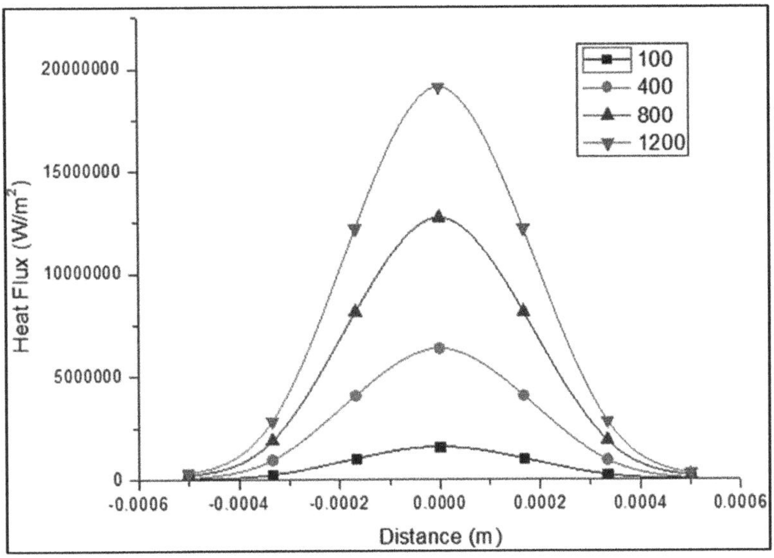

Figure 6.10. *Conduction model for various power inputs. For a color version of this figure, see www.iste.co.uk/kumar/materials.zip*

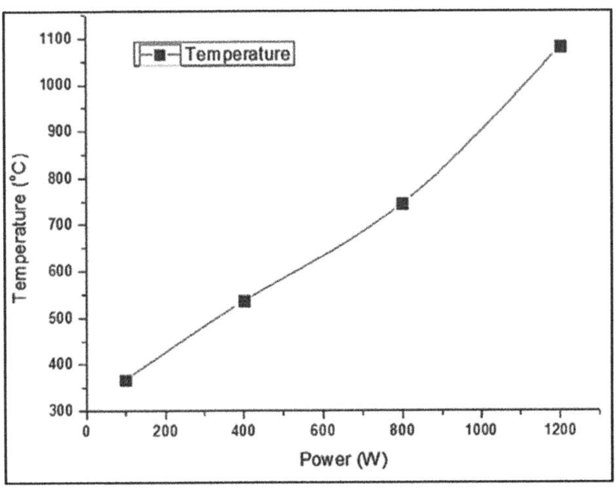

Figure 6.11. *Temperature distribution for various power inputs. For a color version of this figure, see www.iste.co.uk/kumar/materials.zip*

6.4.2.2. *Residual stress analysis*

There is a quadratic reduction in power requirements with the reduction in thickness. The resulting residual stress distribution is shown in Figure 6.12. A designer should note that residual stress remains stable if an appropriate power is used for the selected thickness.

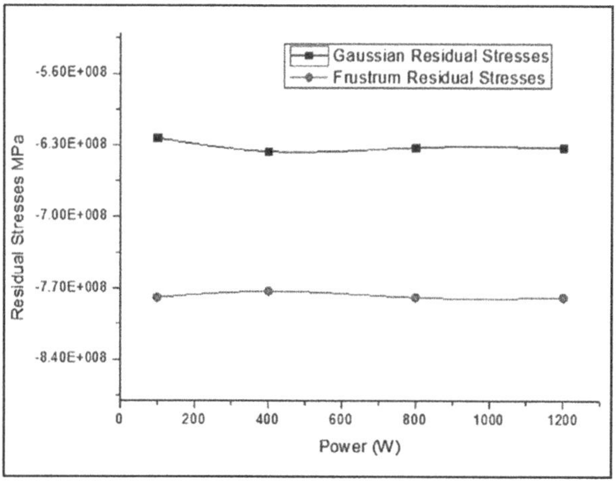

Figure 6.12. *Residual stress distribution for various power inputs. For a color version of this figure, see www.iste.co.uk/kumar/materials.zip*

6.5. Conclusion

A finite element model of the welding process was obtained using the Ansys package. Laser beam welding was substantiated by keyhole formation. A heat flux comparison was made between frustum and Gaussian distributions, and the Gaussian distribution was found to give a better temperature profile in order for fusion to occur.

The heat flow through the keyhole was modeled by a combination of the Gaussian distribution on the surface and a constant heat for the keyhole.

An equal distribution of heat, 600 W from the surface and through the keyhole surface, was taken to calculate both temperature distribution and residual stress.

A study of the Gaussian distribution was conducted, with decreasing bead widths from 4 mm to 1 mm as required in a laser beam. As expected, higher heat intensity with a lower bead width gives higher temperature.

Heat requirements with lower thicknesses can be evaluated by the proper selection of power welding, keeping the residual stress within the yield stress of the material.

6.6. References

Akella, S., Harinadh, V., Krishna, Y., and Buddu, R.K. (2014). A welding simulation of dissimilar materials SS304 and copper. *Procedia Materials Science*, 5, 2440–2449.

Akella, S., Vemanaboina, H., and Kumar Buddu, R. (2016). Effect of heat flow condition in analysis of electron beam welding. *Sreyas International Journal of Scientists and Technocrats*, 1(1), 1–9.

Anupamadev, A.P., Kumar, V.A., Gupta, R.K., Pradhan, S.K., and Kumar, R.R. (2020). Simulation and experimental validation of EBW studies in austenitic stainless steel AISI-321. *Transactions of the Indian National Academy of Engineering*, 5, 581–592.

Belhadj, A., Bessrour, J., Masse, J.-E., Bouhafs, M., and Barrallier, L. (2010). Finite element simulation of magnesium alloys laser beam welding. *Journal of Materials Processing Technology*, 210, 1131–1137.

Fabbro, R. (2010). Melt pool and keyhole behaviour analysis for deep penetration laser welding. *Journal of Physics D: Applied Physics*, 43(44), 445–501.

Goldak, J.A. and Akhlagi, M. (2005). *Computational Welding Mechanics*. Springer, Berlin.

Kaplan, A.J. (1994). A model of deep penetration laser welding based on calculation of the keyhole profile. *Journal of Physics D: Applied Physics*, 27, 1805–1814.

Ki, H., Mohanty, P.S., and Mazumder, J. (2002). Modeling of laser keyhole welding II: Simulation of keyhole evolution, velocity, temperature profile and experimental verification. *Metallurgical and Materials Transactions A*, 33, 1831–1842.

Nath, A.K., Sridhar, R., Ganesh, P., and Kaul, R. (2002). Laser power coupling efficiency in conduction and keyhole welding of austenitic stainless steel. *Sadhana*, 27(Part 3), 383–392.

Rosenthal, D. (1946). The theory of moving sources of heat and its application to metal treatments. *Transactions of the ASME*, 68, 849–865.

Vemanaboina, H., Akella, S., and Buddu, R.K. (2014). Welding process simulation model for temperature and residual stress analysis. *Procedia Materials Science*, 6, 1539–1546.

Vemanaboina, H., Edison, G., Akella, S., and Kumar Buddu, R. (2018a). Thermal analysis simulation for laser butt welding of Inconel625 using FEA. *International Journal of Engineering & Technology*, 7(4.10), 85.

Vemanaboina, H., Edison, G., Akella, S., and Kumar Buddu, R. (2018b). Simulation of hybrid laser-tig welding process using FEA. *Journal of Engineering Science and Technology*, 13(6), 1782–1792.

Vemanaboina, H., Naidu, G.G., Vinod Kumar, G., and Ramachandra Reddy, D. (2019a). Welding characteristics of butt-welded Inconel625 plate using CO2 laser beam. *Materials Today: Proceedings*, 19, 859–863.

Vemanaboina, H., Akella, S., Buddu, R.K., and Gundabattini, E. (2019b). Distortion validation of laser beam welded SS316LN steel plates. *8th International Conference on Modeling Simulation and Applied Optimization (ICMSAO)*, 15–17 April.

Zain-ul-abdein, M., Nélias, D., Jullien, J.-F., and Deloison, D. (2010). Experimental investigation and finite element simulation of laser beam welding induced residual stresses and distortions in thin sheets of AA 6056-T4. *Materials Science and Engineering A*, 527, 3025–3039.

Zeng-Rong, H., Jian-Zhong, Z., Hua-Feng, G., and Jian-Jun, D. (2007). Simulation of temperature field of laser welding by ABAQUS. *Laser Technology*, 31(3), 326–329.

7

Effect of Formability Parameters on Tailor-Welded Blanks of Light Weight Materials

Improvement in today's living standards and enhancement of innovative technology in the automotive manufacturing industry have led industries and researchers to use advanced manufacturing techniques or innovative ideas that discourse customer requirements, as well as design and development ought to be cost-effective and should prove to be beneficial for both the industry and the consumer. Different materials with varying cross-sections need to be used based on requirements in aerospace and automotive industries. To organize this, researchers are enthusiastically proposing a tailor-made welded blank (TWB) strategy, and this technique has been adopted in many automotive industries. This research was carried out to test the formability of TWBs with various light alloy sheets used in aerospace and automotive industries. An overall review of various parameters affecting the formability of TWBs is presented in this chapter, so that other investigators can rely on this for more critical observations in this field.

7.1. Introduction

The automotive industry has paid exceptional attention to arrangements for weight loss and fuel economy over the last few years. Therefore, tailor-welded blanks (TWBs) are mainly used for manufacturing the automotive

Chapter written by Dappu DEEPIKA, Akkireddy Anitha LAKSHMI, Tanya BUDDI and Chalamalasetti Srinivas RAO.

outer body. A TWB is prepared from several different materials that are combined with similar or dissimilar strengths or thicknesses to shape a sole component before the sheet is formed. The key benefit of consuming a TWB is experienced at crucial sections of the sheet metal blank, which gives thicker or more grounded materials to build up the stiffness of the component. It also benefits by reducing the heaviness of car outer bodies. Using the finite element analysis, Padmanabhan *et al.* (2008) manufactured TWBs with the grouping of aluminum and stainless steel sheets and the formability characteristics of welded blanks. Panda *et al.* (2009) conducted formability tests of welded sheets in the forming phase of a plane-strain stretch. In order to forecast the formability variation between without weld and tailor-welded blanks, Ganesh Narayanan and Naik (2010) framed new failure principles. Dilmec *et al.* (2013) experimented and produced an AA2024 aluminum alloy forming limit diagram and investigated the role of sheet thickness and anisotropy properties in welded blanks. Palanivel *et al.* (2011) demonstrated that the to and fro action of the pin(square) could fruitfully weld dissimilar welded aluminum blanks using the FS-welded method and the pin profile (square) tool to achieve good results, compared with other tool pin profiles. An experimental analysis was conducted by Choi *et al.* (2000) using TWBs with two thickness combinations (0.8t×1.2t and 0.8t×1.6t) and using SPC1 sheets. Chen *et al.* (1996) developed a clear relationship between ductility of the weld joint and the cup height for vertically joined TWBs by stretching under the plane strain condition that constitutes parent metals of similar thickness, metal and combination of covering. Scriven *et al.* (1996) used the limit dome height (LDH) test to evaluate the formability of TWBs. In this test, specimens (101.6 mm×177.8 mm) were intended to produce a level strain condition, which represented both the conditions of the lowest failure strains and the strain state in which most of the automotive panels failed. Kridli and Kusuda (Kusuda *et al.* 1997; Kridli *et al.* 2000) mainly considered the impact of the TWB welding line direction on the incorrect minimum major strain levels resulting from the forming limit diagrams (FLDs). These researchers further pointed out that the formability of TWB was significantly influenced by the thickness of the thinner material rather than the thickness ratio. However, they did not provide further information on the strain distributed in the FLD during the analysis. (Chen *et al.* 1997) indicated that the FLD intercept on the main strain axis was an element of the sheet thickness. Thus, most of the researchers have only focused on the effect of sheet thickness on the base metal, as opposed to a comparative study on TWBs.

7.2. Experimental procedure

N. Bhanodaya Kiran Babu *et al.* (2014) showed the formability of TWBs of two materials (AA6061 and AA2014). The composition of the compounds AA6061-T6 and AA2014-T6 is given in Table 7.1, and their properties are presented in Table 7.2. Examinations were carried out according to an experimental plan of the Taguchi L9 orthogonal array. AA6061 was used in the T6 state, and AA2014 was treated with a solution. Five days after preparing the solution of AA2014 samples, TWBs were made because of the aging of the compound hardened by the precipitation process, having strength less than the AA series 6061 material from three to five days. The tensile test carried out on the blanks showed that the elongation percentages of AA6061 and AA2014 were 22% and 12% respectively. The aluminum alloy AA series 2104 is solutionized at 495°C for 1 hour, in an electric furnace, without considering the pre-heating time of the furnace before the FSWed, in order for the blank to distort equally. The three information parameters of the orthogonal Taguchi array and welding were implemented using three levels. By changing the three variables into three factors, the process parameters were obtained from nine combinations.

Elements	Mg	Si	Cu	Ti	Cr	Al
6061-T6	0.60	0.76	0.025	0.017	0.043	Balance
2014-T6	0.46	0.75	3.98	0.026	0.017	Balance

Table 7.1. *Elemental composition*

Base Material	Yield strength (MPa)	Tensile strength (MPa)	Percentage elongation (%)	Micro hardness (Hv)
AA6061	198	273	22	115
AA2014	402	454	12	165

Table 7.2. *Properties of standard materials*

The appar. atus had a shoulder diameter of 20 mm, a pin profile of 6 mm with a threaded taper and a pin length of 2.5 mm. As a recreated plane strain condition of 100 mm×60 mm blank size, rectangular specimens of 100 mm×60 mm were cut from the welded sheets and produced at a punch distance of 36 mm. The LDH test was performed at a punch speed of 0.3 mm/s on a 50 ton hydraulic press

N. Bhanodaya Kiran Babu *et al.* (2015) considered the TWBs joined by FSW with five various pin profiles of tools. The formability of welded blanks was assessed at constant parameters of the friction-stir welding process. In this present study, rolled aluminum alloy plates, namely AA series 6061-condition T6 and AA series 2014-condition T6, were used as standard alloys. Table 7.1 shows the chemical composition of the base materials, and Table 7.2 presents their mechanical properties. The process parameters used were the turning speed of the tool of 900 rev/min, the weld of 4 mm/in (speed) and the tool angle (inclination) of 1 degree. Figure 7.1 shows different tool pin profiles for FSW, namely taper cylindrical (TC), straight square (SS), stepped cylindrical (ST), straight hexagon (SH) and straight cylindrical (SC).

Figure 7.1. *Photographs of the pin profiles of different tools according to N. Bhanodaya Kiran Babu* et al. *For a color version of this figure, see www.iste.co.uk/kumar/materials.zip*

Microhardness tests were conducted on the cross-segment blanks of the weld, and the values were estimated over the cross-section of the weld using the Vickers microhardness analyzer with a load of 200 g for a dwell time of 10 s. The formability of welded joints was investigated by the LDH test at a punch speed of 0.3 mm/s and hemispheric punch on press (hydraulic) of 50 ton. The square (100 mm × 100 mm) tests for biaxial conditions were prepared from welded blanks and extended with a punch to the fracture. The scanning electron microscope (SEM) examined the microstructure of the welded blanks. To uncover the microstructure, samples were cleaned with various different emery papers (various grades), and etching (Keller's reagent) was carried out.

Arman Khan *et al.* (2014) considered two different steels with the impact of aluminum TWBs by changing the ratio of thickness on the characteristics of formability. The blank sheet of aluminum (AA6016-T4) was combined with two high-quality steels (HSLA-340 and DP600) with three ratios of thicknesses (steel/Al) of 1.0, 1.25 and 1.5 to form six dissimilar combinations. Simulation of stretch forming of parent materials was performed using the LS-Dyna software.

In this work, aluminum (AA6016-T4) was combined with Al-steel TWBs, DP600 and HSLA-340. For each combination of materials, three combinations of thicknesses, namely 1.0t*1.0 t, 1.25t*1.0 t and 1.5t*1.0 t, were used to verify the data on the thickness effect. Steel tend to diffuse less in the joint nugget than aluminum due to temperature of insufficient and time interface of the weld; thus, a centralization of 40% for steel and 60% for aluminum was used. In the present examination, the blank holder was marginally divided into equal parts to apply dissimilar forces on the two TWB sections. The material stream was controlled by this methodology depending on the strength of sheet segments, so that it did not cause wrinkle nor tear. A 5 KN force was applied to the blank aluminum, and 3 KN was applied to the blank steel.

Vijay Gautam *et al.* (2019) tested dual-phased (DP) steel of grade TWBlank ranged between 1.4 and 1.8 mm. TWBlank ranged between 1.4 and 1.8 mm. The dual-phase steels produced through annealing resulted in the formation of a microstructure containing austenite phase and ferrite. The ASTM E8 M method was used to verify the tensile properties on a 50 kN UTM in uniaxial tension of the weld area, TWBs and parent sheets. For example, the YS (yield strength), TS (tensile strength), elongation percentage, hardening model, anisotropic coefficients and the strength factor of the specimens were examined at room temperature to evaluate significant properties. Distinctive true stress and true strain curves plotted using the data, and the tensile tests obtained are shown in Figure 7.2. The tensile tests are in accordance with E512 ASTM standards for the experiment arranged at 0°, 45° and 90° to the route of rolling of blank in order to determine the typical anisotropic coefficients. Anisotropy tests were conducted using a uniaxial testing machine extending by applying a 15% load.

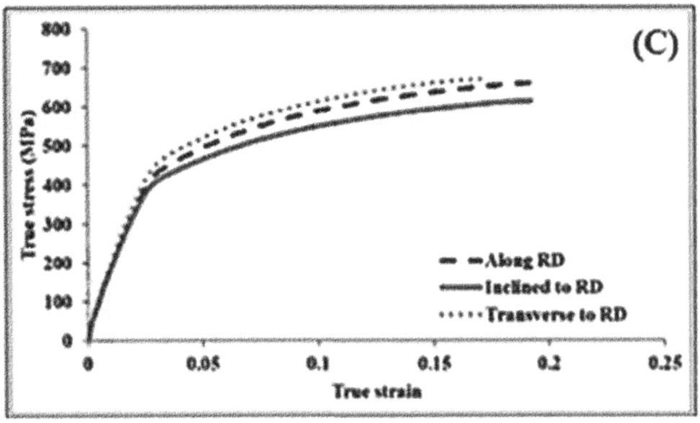

Figure 7.2. *True stress–true strain curves of 1.4 mm-thick JSC590RN steel samples with different rolling direction orientations according to Vijay Gautam* et al. *For a color version of this figure, see www.iste.co.uk/kumar/materials.zip*

The LDH test assesses the formability of a sheet metal. The test incorporates the ASTM standard biaxial direction of the sheet metal stretching over a hemispheric punch. A blank of 200×200 mm set apart with grid circles of diameter 2.5 mm was verified between a holder blank, die and a punch using a 100 ton hydraulic press, with data acquisition system displacement being given downward until sheet necks as shown in Figure 7.3. Different parts such as die punch and blank holder is shown alongside the deformable blank using the ABAQUS FEA software to complete the simulation. The outside diameter, the inner diameter and the die of 255 mm, 104 mm and 35 mm in depth were respectively the different die dimensions used in the simulations.

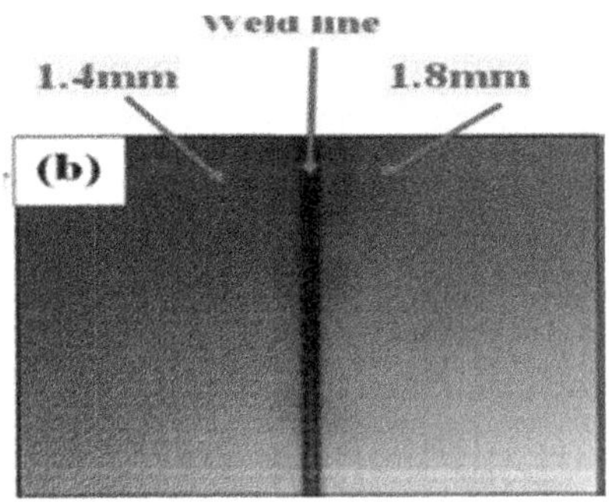

Figure 7.3. *Grid marking of the TWB according to Vijay Gautam* et al. *For a color version of this figure, see www.iste.co.uk/kumar/materials.zip*

M. Habibi *et al.* (2018) tentatively examined the properties (material), forming, fracture forming limit diagrams (FLDs) of tailor spaces (TW Blanks) delivered by friction-stir welded (FSW). They used mild St14 steel. To check the mechanical properties of the sheet, the uniaxial tensile test, as indicated by regular ASTM E08, was used. Welding test sheets were created in 250×55×1.4 and 250×55×1.2 respectively. Different rotational and feed speeds were used for the FSW process. Three rotational speeds (800, 1100 and 1400 rpm) and three feed speeds (60, 100 and 140 mm/min) were selected where it was found that ideal speeds attempt are error prone. A semi-automatic machine (milling) is used to perform the welding procedure. Four types of tailor blanks were formed: the first type was with a similar thickness and a rolling welding direction; the second type involved a similar thickness in a cross-sectional direction; the third type included dissimilar thicknesses (1.2 and 1.4 mm); and the fourth type was created using the grinding procedure.

S.M. Chan *et al.* (2003) selected SPCC (cold-moved) steel sheets as the experimental material. The sheets with a thickness of 0.5, 0.6, 0.8 and 1.0 mm were used for butt laser welding and cut into various dimensions. All tailor specimens were divided into two regular forms, with breadths ranging from 20 to 152 mm. The sequence rounded blanks were used in the tests. A laser ND: YAG was used for butt welding the tailor-made blanks, as shown in Figure 7.4. To test the formability of TWB progression, the swift round-bottom test was used, with a round punch diameter of 50 mm. Throughout the investigation, the plain holding strength (30 kN) was kept constant. In the middle of the specimen, the apparatus of dissimilar test specimens (tensile) was cut with the joint line, which was right angled to the tensile test axis. The machine used was the MTS tensile test and all TWBs were steered at a constant cross-head speed of 2 mm/min until the necking occurred.

V. Paradiso *et al.* (2017) showed the FSW connection between the dissimilar Mg alloy and the AA series 2024-T3 good-quality composite aluminum. He focused on the practicality of the FSW procedure to yield better joints. The 4 mm-thick sheets AA2024-T3 and ZE41A were used as standard materials. AA2024-T3 was provided as rolled treated sheets. In the forward side of the butt joint, the ZE41A sheet was fixed to reduce the welding heat input and to decrease the brittle IMC arrangement as much as possible. Friction-stir welding was performed, using a machining center (MCX 600 ECO) to offset the 1 mm tool to the magnesium side. The tool specifications consisted of a 20 mm shoulder diameter funnel-shaped unthreaded pin (3.80 mm high, 6.20 mm diameter and 30° cone angle). The accompanying processing FSW parameters were rotational speeds, feed rate, tilt edge and shoulder plunge depth. After plunging, dwelling and welding, the FSW procedure was applied. The plunging step was conducted at a feed rate of 5 mm/min, while 10 sec was set as the dwell duration. The welded specimens were inspected for joint microstructure and hardness at the joint. The specimens were placed in a legitimate thermoset resin and ground by the tissue disk using different grade abrasive disks (and later cleaned with a suspension made of polycrystalline diamond until the surface had a mirror-like finishing). The test specimens were etched chemically by a reagent to reveal the critical highlights of the microstructure in aluminum. A similar system was used to find the magnesium surface using a picric acid solution in a blend of ethanol, acetic acid and refined water.

Figure 7.4. *Tailored Nakazima specimens according to S.M. Chan* et al.

Sameer MD *et al.* (2019) used base metals (BM) such as AZ91 magnesium alloy and AA series 6082-tempered T6 aluminum alloy sheets of dimensions 220×75×6 mm, which are industrially available. The samples were cleaned by emery paper, in order to remove the dirt and rust layer before welding. Welding was carried out by a milling machine with three axes. It used a smaller cylindrical threaded tool made from H13 device steel. The specifications of the FSW device are 18 mm shoulder diameter with a 3 mm pin diameter and a 5.6 mm pin length. The device was rotated in a counterclockwise direction and the tilt angle of the tool (1.5°) was kept constant. The overall parameters were varied to observe the impact of the material stream on the AZ91 Mg alloy and AA6082-T6 Al FSW. At a constant tool rotational speed of 560 rpm and a tool traverse speed of

36 mm/min, four unique joints were installed. The joints were characterized as AS, S2, D2, S1 and D1, depending on the type of joint AA series 6082-tempered T6 Al alloy when the AZ91 Mg alloy was on the advancing side. Using a visual microscope and electron microscopy, the progression of the alloy, the structure of the grain and the analysis of the joint between faces were observed after the tensile test.

Y.J. Kwon *et al.* (2008) performed dissimilar FSW of A series 5052P-O, AZ magnesium 31B-O alloy sheets. They tentatively explored the impacts of the rotation of the tool on the surface presence, grain orientation and tensile properties of the friction-stir (FSWed) sheets. The dimensions of the base metal plate were 30 mm broad, about 160 mm long and about 2 mm thick. Throughout the FSW process, AZ magnesium 31B-O alloy plates, A series 5052P-O were positioned at the retreating side and the forward side respectively. For FSW, the radius of the tool shoulder was 5 mm, and the probe was 4 mm wide and 1.7 mm tall. The tool was rotated at speeds ranging from 800 to 1,600 rpm in a clockwise direction, and the probe was then embedded into the materials. At a constant speed of 300 mm/min, the rotating tool was crossed along the joint line in the perpendicular direction of the rolling of base metal plates. The FSW plates were perpendicularly cut along the traverse direction of the tool for the tensile test, and subsequently machined into the tensile test specimens with a parallel portion of 4 mm broad and a fillet radius of 5 mm. Tensile testing was performed at room temperature at a steady crosshead speed of 1 mm/min.

Aruri Devaraju *et al.* (2019) investigated the impact of the tool on the tensile and microhardness properties of dissimilar 2024 and 6061 aluminum alloys obtained using the friction-stir welding process. The investigation was conducted on a 6 mm-thick sheet 6061 base metal (Al-Mg-Si compound) and aluminum compounds 2024 (Al-Cu mixture). The plates were joined by assembling two plates using the FSW process on a vertical milling machine. The pin tip of the dowel is forced into AA2024–AA6061 specimens between 180 and 300 cycles for every second. The pin tip of the dowel was constrained under 5,000–10,000 pounds per square inch of force. For all of the FSW tests, a 5 KN force was applied and the instrument was tilted to 1.5°. The material selected for the instrument was H13 tool steel. Tests were performed at three different tool rotational speeds of 900, 1,120 and 1,400 rpm and a constant welding speed of 40 mm/min. A microstructural study was performed after FSW by cleaning the samples with a disk cleaning machine using Keller's reagent and viewed under an optical microscope

(OM). The tool design provides the consistency of microstructure and properties in the same way as the procedural loads. The characteristics of the fracture were revealed by the tensile test.

Sundaram *et al.* (2010) assessed the impact of process parameters, such as pin dimensions of the tool, pin speed, tool speed and slope inclination, on the tensile strength of the base materials AA series 6061- condition T6 alloy and AA series 6082 alloy plates. Metal plates were degreased and air dried from acetone. These plates were butt joined and arranged using the FSW process along the longitudinal direction. For example, two tool pin square (4 mm × 4 mm) and round (measurement 6 mm) pin height (4 mm) were used to stir the alloy during the FSW method. The shoulder width of the plane was fixed as 16 mm. H11 welding material tools were used. The trials were conducted on an FSW-altered milling machine (Model-FN2V). The specimens were sliced into regular sizes using a wire-cutting electric discharge machine. The tensile specimens were sliced perpendicular to the welding joint, according to the standard ASTM E8.

H.M. Anil Kumar *et al.* (2020) explored the tensile strength of the weld under three distinct estimates of shoulder diameter, offset of the tool and tilt angle of the tool. The 5083 and 6082 aluminum alloys on the Vickers hardness scale had individual hardnesses of 85 and 95 respectively. The two alloys were thoroughly machined by the milling machine to obtain a measurement of 140 ×70 ×4 mm, and the work top surfaces were rubbed with emery paper. The pin profile of the square tool taper made of HSS having 3 diameters of shoulder range (16–20 mm) is presented in Figure 7.1. The tool pin had a constant length of 3.4 mm and a breadth of 3 mm on the top and 5 mm on the square's base substance was intended for the dimension of the tool pin. In the study by Anil Kumar *et al.* (2018), the procedure parameters were kept constant but the tool parameters were changed.

K. Palani *et al.* (2018) investigated the impact of pin (profile) and speed of rotation on the micrograph and the strength of unique tensile AA series 8011 and AA series 6061-T6 aluminum friction-stir-welded alloys. Friction stir welding tests performed with aluminum alloy plates (150×75×6 mm) that were butt joined and AA8011 cold worked and AA6061-T6 heat treated were cut using sawing and milling-type machines. Hardness tests were carried out on a Vickers microhardness analyzer (Shimadzu, HMV 2 T E) using a 2 kgf load with a 15 s holding time, based on the requirements of ASTM E92. The hardness values were estimated from a separation of

0.5 mm between the estimates at the center of the weld along the cross-section perpendicular to the welding direction. The cross-segments of the welded joints were polished using various grade numbers of emery sheets, and then cleaned and polished with a diamond paste and an alumina paste respectively. The specimens were then etched using the Keller reagent (Sharma *et al.* 2012). The micrograph of specimen weld was examined using a visual (Olympus) microscope and a (BRUKER 1000) SE microscope, and ED spectroscopy was used to locate the elementary properties of the welded joints.

7.3. Results and discussion

N. Bhanodaya Kiran Babu *et al.* (2014) estimated the mechanical and non-treated formable properties and welded blank (heat treated). The limiting dome heights of the base materials AA6061 and AA series 2014 were 19.4 mm and 6.2 mm respectively. The LDH estimate for the base material heat treated with AA2014 was 12.4 mm, indicating formability that can be attained by heat (treatment). The values obtained for the LDH were plotted against the rotational speed, the weld speed and the tilt angle. The formability response was influenced by the process parameters. The tool rotational speed was increased from 900 to 1,400 rpm, resulting in a decrease in formability. The frictional force on the blanks increased the rotational speed of the tool, increasing the input heat. The heat input improved grain development, but led to poor formability ranging from 900 to 1,400 rpm after the speed was increased. When the welding speed was adjusted from 16 to 24 mm/min, it resulted in increased formability, while that adjusted from 24 to 32 mm/min resulted in reduced formability. Similarly, the change in the tilt angle of the tool from 0° to 1° increased formability, and that from 1° to 2° decreased formability. This resulted in the superior mixing of joined metals and improved formability at a tilt angle of the tool of 1°. Despite this, when the tilt angle of the tool was increased to 2°, it resulted in poor formability, which was attributed to the influence of metal cohesion due to the increased development in the vertical direction of the stirred zone.

Figure 7.5(a,b) shows the microstructure of the AA6061 and AA2014 materials. The AA6061 microstructure comprises a small number of Mg_2Si precipitates, and the structure of the AA2014 aluminum alloy comprises immiscible iron and aluminum particles. The normal aging (heat treatment) AA2014 sample found CuAl2 fine precipitates throughout the network (Figure 7.5c). Figure 7.5(d–f) shows the micrographs of the dissimilar welded plates under variable welding conditions. Microstructures show both AA2014 and AA6061 alloys under all the welding cases because of the mixing (stir) action. Figure 7.5(d) and (e) presents the homogeneous mixing of the constituents occurring at 900 and 1,150 rpm with fine grain. Figure 7.5(f) shows that the alloy is welded with AA6061 and AA2014 patches, resulting in lower formability.

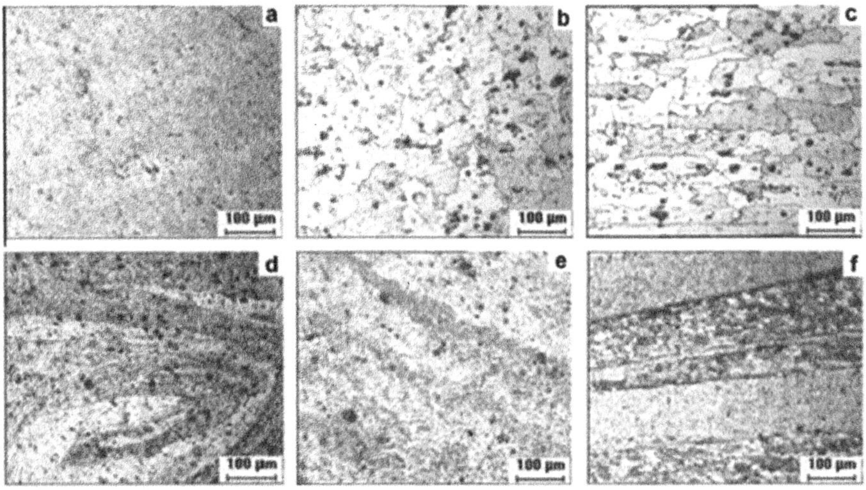

Figure 7.5. *Micrographs (weldable) of TWBs: (a) AA6061, (b) AA2014, (c) naturally aged AA2014 alloy, (d) TWB 900 rpm, (e) TWB 1,150 rpm, (f) TWB 1,400 rpm, according to N. Bhanodaya Kiran Babu* et al.

N. Bhanodaya Kiran Babu *et al.* (2015) studied the microhardness distribution of welded blanks composed of dissimilar profiles of tool pins, which was introduced on the cross section of the weld zone (Figure 7.6). It can be clearly seen that the durability of the weld area is less than the standard materials for welded blanks composed of different profiles of tool pins. The welded zone is the softened region of a welded blank, where degradation of hardness occurs (Palanivel and Koshy Mathews 2011). Of the selected five blanks (welded), the blank with the profiled pin tool (SS) had high hardness.

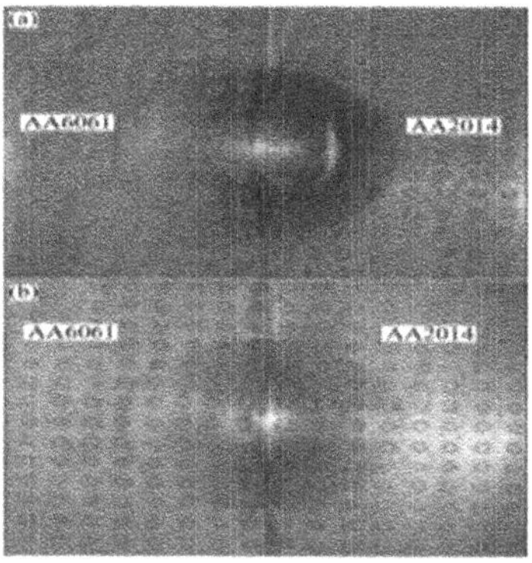

Figure 7.6. *Deformed blanks made up of aluminum and SS alloys according to N. Bhanodaya Kiran Babu* et al.

SEM graphs of the blanks of weld stir zones made of dissimilar pin profiles of the tools are shown in Figure 7.7(a–e). The stir area microstructures of pin tools (TC, ST) are not homogeneous compared to the SS tool, which results in reduced formability. The stir area microstructures of profiled tools (SH, SC) are also not homogeneous and have rough grains due to a lack of heat input.

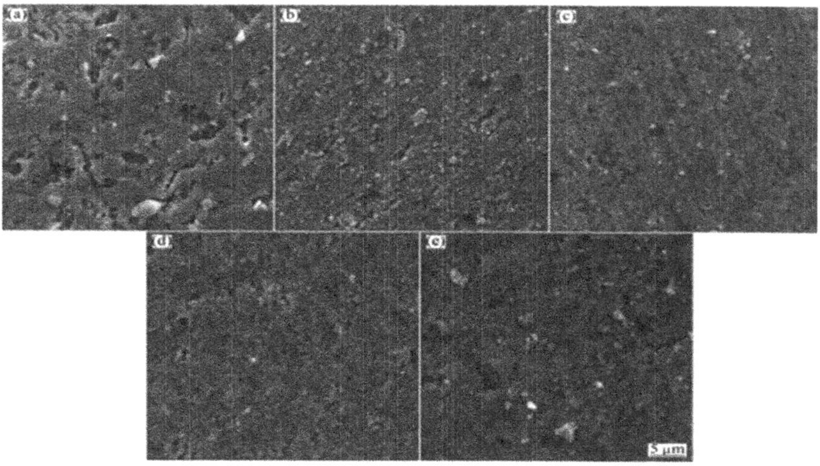

Figure 7.7. *SEM graphs of the welded zones made of the pin profiles of the SC tool, according to N. Bhanodaya Kiran Babu* et al.

Arman Khan *et al.* (2014) demonstrated that the weld line moved in the deep drawing of TWBs due to material mismatch. As a result of bigger material flow in aluminum, in the cup section, the weld line moves towards the steel side; however, in the flange part, it moves towards the aluminum side, as shown in Figure 7.8. Huge weld line movements are seen, contrasting with Al-HSLA combination in the Al-DP combination. By varying blank holder forces on the TW blank aluminum-steel, the movement of the weld line is controlled. As shown in Figures 7.8 and 7.9, the movement of the weld line tends to increase as the proportion of thickness is increased. This is because when compared to the thinner steel, the material flow of the thicker steel is less. The variation in thickness along the diagonal direction for every ratio of thickness is limited by optimizing the shape of the TWB before the deep drawing operation. For all of the three thickness variations, compared to Al-DP, there was additional material flow in the Al-HSLA combination. In all of the cases, the pattern of the variation in thickness was the same. As the thickness ratio increased, the minimum thickness value also increased in the TWB.

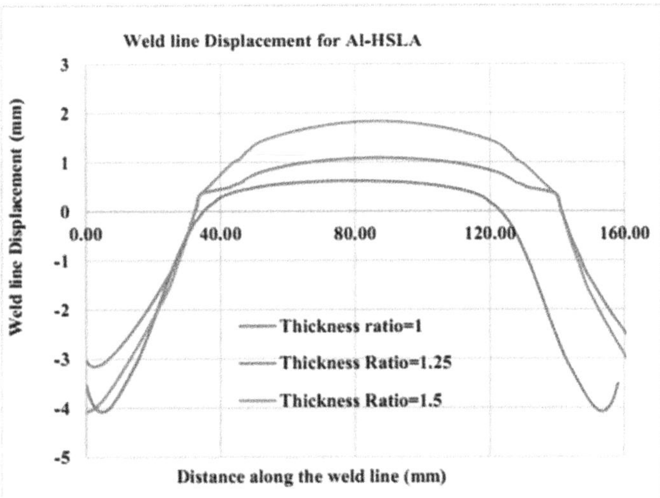

Figure 7.8. *Al-HSLA welded line moves for three thickness ratios according to Arman Khan* et al. *For a color version of this figure, see www.iste.co.uk/kumar/materials.zip*

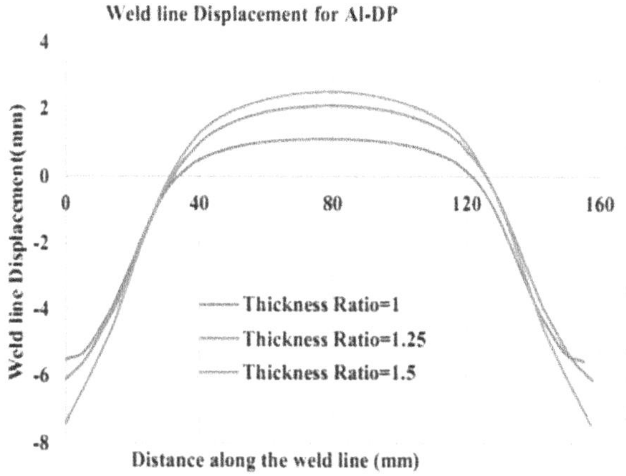

Figure 7.9. *Welded Al-DP line moves for three thickness ratios according to Arman Khan* et al. *For a color version of this figure, see www.iste.co.uk/kumar/materials.zip*

Vijay Gautam *et al.* (2019) showed that the average YS (yield strength) for a thick sheet of 1.4 mm was about 421 MPa, and for a thick sheet of 1.8 mm, it was 415 MPa. The tensile strength for a 1.8 mm-thick sheet was slightly higher than that for a 1.4 mm-thick sheet. The percentage elongation was approximately 23% for both sheets. It was observed that the elongation percentage for 1.4 mm-thick sheet tensile specimens oriented at 0° to the rolling direction was more than that for the other two route directions. Similar observations were found in 1.8 mm-thick blanks. The strain hardening exponent for the thicker sheet was greater than that for the thinner sheet. The 0.2% offset yield strength of the weld zone was 770 MPa, which was double the yield strength of the individual sheet. For example, the tensile strength of the weld zone was 883 MPa, which was 1.5 times more than that of the parent sheets.

M. Habibi *et al.* (2018) selected three rotational speeds and three feed speeds to extract optimized parameters, and assessed the welded samples after performing the welding procedure. They found that four parameters had undesirable appearances. Microscopic tests were made to examine the microstructural and metallurgical phenomenon in the joint area. As shown in Figure 7.10, the variable 1,100 rpm and 60 mm/min was most valid from microstructural and appearance points of view, and the parameters present in grains (1,400 rpm) were coarser and broader because the temperature expanded everything that might be needed. The tensile test on both weld method tests across the weld length was performed to precisely make the best decision between the two speeds (1,400 and 1,100 rpm).

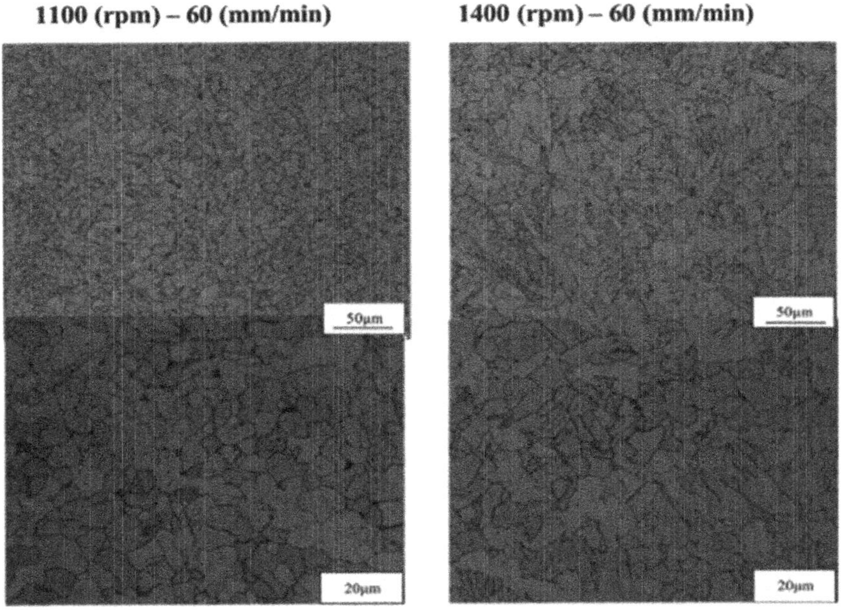

Figure 7.10. *Metallographic images from the welded sample center, according to M. Habibi* et al.

Figure 7.11 shows that the formability decreases with welding. As shown in Figure 7.12, welding decreases the formability of tailor-made blanks. Similarly, Figures 7.12 and 7.13 indicate that the differences in thickness and welding have a negative impact on FLD.

S.M. Chan *et al.* (2003) demonstrated the use of swift round-base tests to examine the formability of TWB. Figure 7.14 shows the type of failed TWBs. The failure of TWB corresponded to the weld line, and was perpendicular to the main strain. It occurred at the thinner TWB piece. A thinner material for the most part opposes less force compared to a thicker material. Deformation of the elastic curve is thick when the thinner material experiences plastic deformation. TWBs fail at the region of less thickness

(Sigeert and Knabe 1995). Likewise, in Figure 7.14, the deformed section grids near the fracture region are clearly visible. Currently, deformed circles are estimated on the necking area and those near the crack region. In the test, it was observed that for the purpose of cracking, it was somewhat difficult to control at the necking area. Subsequently, necking and cracking under both uniaxial and biaxial conditions were considered as the cause of failure.

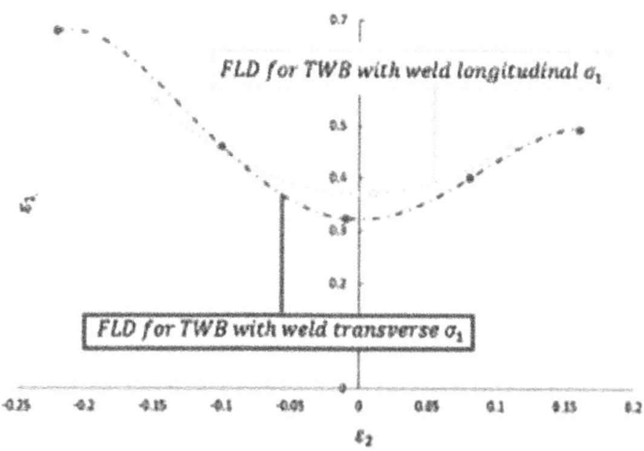

Figure 7.11. *Investigation of the weld in the transverse direction of similar thickness, according to M. Habibi* et al. *For a color version of this figure, see www.iste.co.uk/kumar/materials.zip*

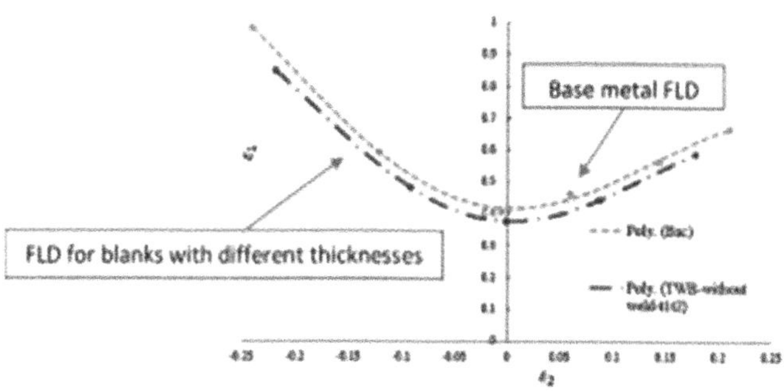

Figure 7.12. *Effect of the change in thickness by equating standard alloys and dissimilar blank sizes, according to M. Habibi* et al. *For a color version of this figure, see www.iste.co.uk/kumar/materials.zip*

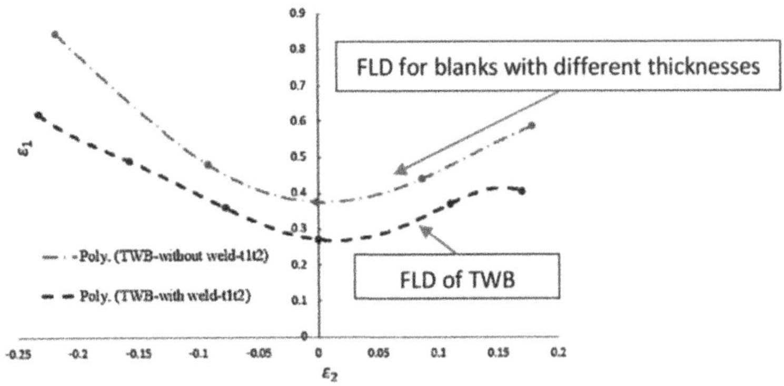

Figure 7.13. *FLD developed by dissimilar weld blanks, according to M. Habibi* et al. *For a color version of this figure, see www.iste.co.uk/kumar/materials.zip*

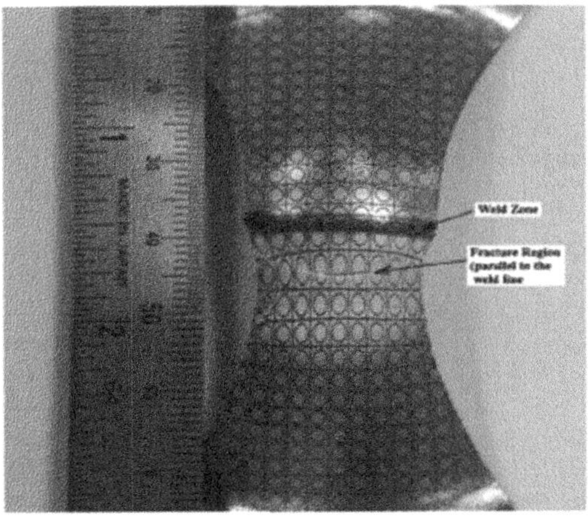

Figure 7.14. *TWB failure type according to S.M. Chan* et al.

V. Paradiso *et al.* (2017) showed that the ZE41A Mg alloy had advanced brittleness compared to the Al alloy, and along these lines, plastic deformation was lower in ZE41A than in AA2024-T3. Consequently, the joint development between the composites AA2024-T3 and ZE41A was inherently unpredictable. The AA series 2024-tempered T3 composite had a nearly double thermal conductivity of the ZE41A magnesium alloy in our study. Because of this considerable dissimilarity, heat dispersion generated throughout the AA2024-T3 and ZE41A welding process was non-uniform, thereby developing high thermal stress levels. If these thermal stresses are not balanced, they may cause hot cracking at a tool rotational speed of 1,400 rpm and a feed rate of 20 mm/min, as observed in the joint handled (Figure 7.15.a). Crack initiation is supported by the proximity of the Al matrix scattered intermetallic compounds (light area) inside the stirred zone, as shown in Figure 7.15(a), from a morphological perspective. It shows that the measurement of heat generation in the ZE41A alloy for a similar tool rotational speed and feed rate differs from the AA series 2024-tempered T3 alloy, and subsequently developing a wide range of values to reduce the initiation of cracks in hot conditions is a vital undertaking to achieve better metallurgical consistency. Weak joints for each feed rate state can be observed at 1,000 rpm (Figure 7.15b).

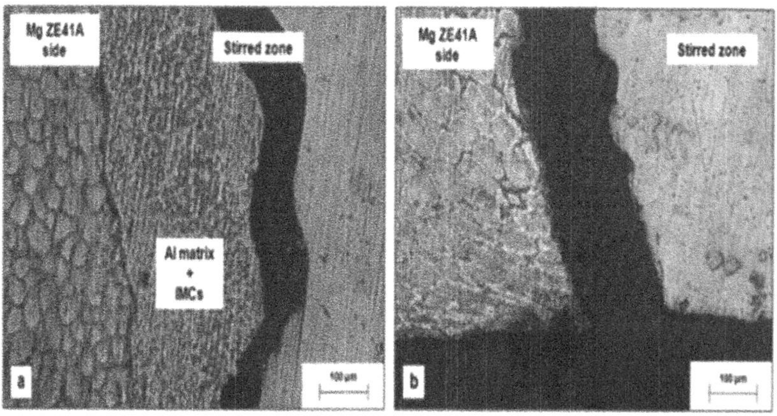

Figure 7.15. *Defects observed: (a) hot cracks inside the stirred zone; (b) view of a weak joint, according to V. Paradiso* et al.

As shown in Figure 7.16, an unpredictable stream design occurs in the mixed zone and an enormous unique growth of new crystals can be seen in the weld and progress areas, with a definite reduction in the dimensions of grain from starting materials to the conversion zone to the welded area. In the FSW joint stir zone, 16 fine dispersed aluminum IMCs can be observed with an exceptionally rich matrix. The mixture of fine grains and smaller particles of IMCs inside the mixed zone can be seen in Figure 7.16. Methods for microhardness measurements are affirmed. Again, a minute reduction beneath the hardness value of standard materials is detected promptly past the mixed zone, all the more decisively in the correspondence of both

Al 2024-T3 and Mg ZE41A sides with the heat-affected zone (HAZ). This is legitimately identified in the way that hardness decreases with expanding grain size in the HAZ, as described by the Hall–Petch condition.

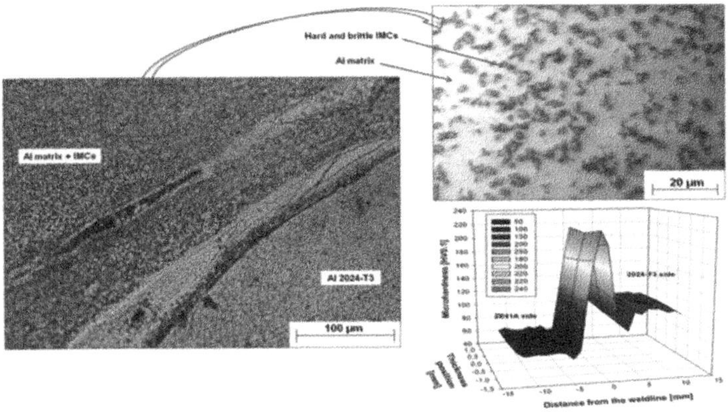

Figure 7.16. *Measures of microstructural and microhardness joints, according to V. Paradiso* et al. *For a color version of this figure, see www.iste.co.uk/kumar/materials.zip*

Sameer MD *et al.* (2019) showed that the Al/Mg contact was observed when Mg was on the advancing side (Figure 7.17(a)). Figure 7.17(c) clearly shows brittle cracks by forming a line, gaps and intermetallic compound layers. A good joint with a better microstructure can be seen in Figure 7.17(b). The interface between the weld zones is displayed in Figure 7.17(c) with inner details containing stage Mg17Al12 limit as α+β and α-phase at the aluminum alloy stage. The stirred zone more closely monitors the stage Mg17Al12 (shown by the arrows) and decreased α-grains (secondary phase) of a reduced grain dimensions (shown by the black arrows), similar to the results reported by Prasad *et al.* (2018). Figure 7.17(d) shows rougher grains at the limits, with a small number of Mg17Al12. In the stir zone of D1, the higher aluminum measure is dissolved along these lines from the observations of the microstructure, which suggests better mechanical properties than the D2 specimen. Subsequently, it tends to reflect that when Mg is applied to AS, aluminum content in the lump zone becomes more soluble than Mg when applied to RS, affecting the mechanical properties.

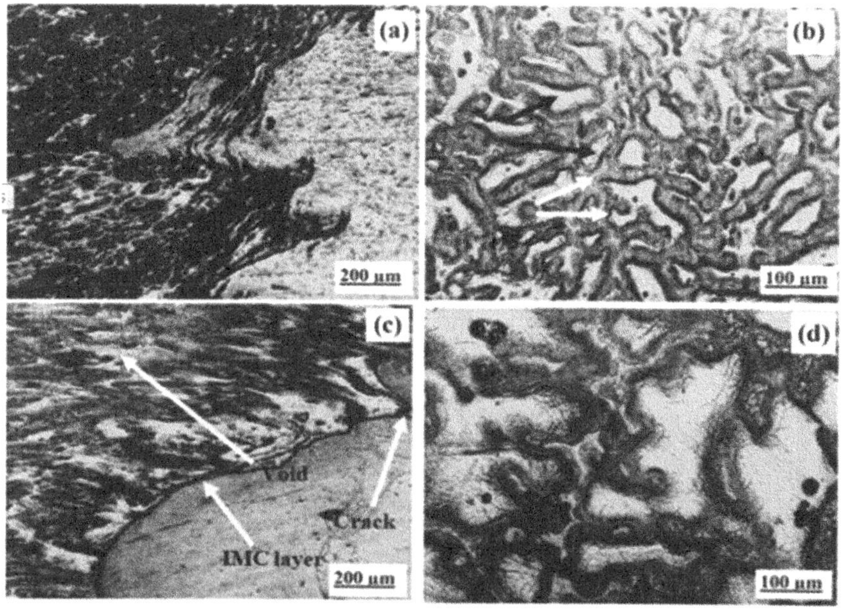

Figure 7.17. *Micrographs showing (a) the D1 mixture pattern, (b) D1 at the stir zone, (c) the D2 mixture pattern and (d) D2 at the stir zone, according to Sameer MD* et al. *For a color version of this figure, see www.iste.co.uk/kumar/materials.zip*

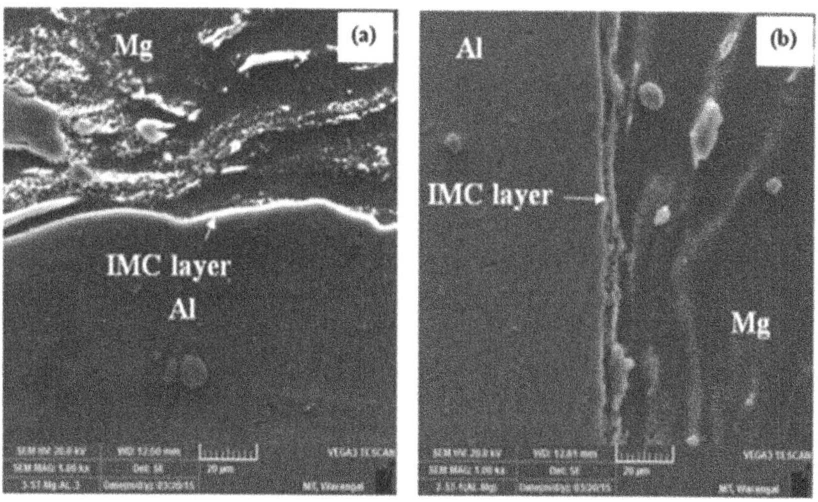

Figure 7.18. *SEM images of weld contacts of different alloys: (a) D1 and (b) D2, according to Sameer MD* et al.

Figure 7.18 presents the SEM images of the enlarged weld at contact. It reveals a small intermetallic layer in the D1 specimen's Al/Mg joint interface when, on the forward side, the Mg alloy is placed as shown in Figure 7.18(a). For the specimen D2 interface, the thick intermetallic layer with weak contact of aluminum with magnesium can be observed in Figure 7.18(b).

Figure 7.19 presents fractography images of comparable aluminum 6082 alloy-T6 welds and dissimilar Mg–Al welded specimens when set to AS. Ductile crack is shown in Figure 7.19(a,b), where the S1 weld shows dimples on the surface of the fracture.

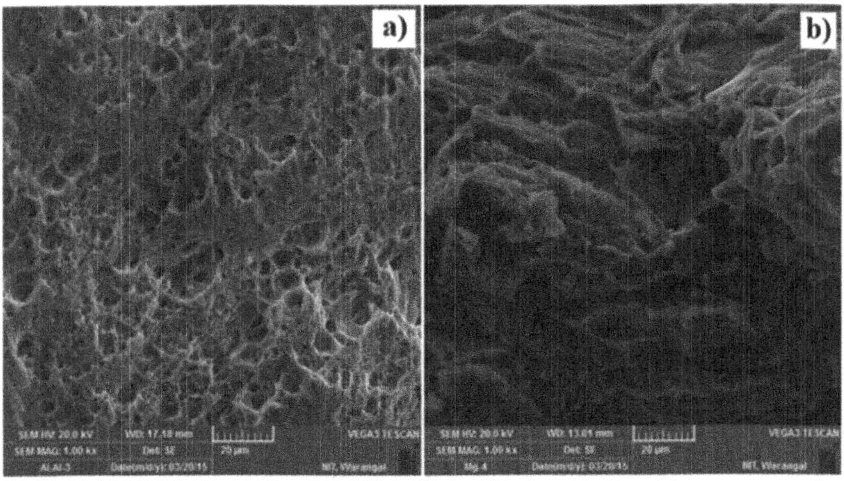

Figure 7.19. *Fractography of the tensile test specimen: (a) Al–Al type weld and (b) dissimilar Mg–Al type weld, according to Sameer MD* et al.

Y.J. Kwon *et al.* (2008) investigated the micrographs of FSWed sheets along the tool movement direction (Figure 7.20(a)–(c)). Yan *et al.* (2005) revealed that cracks were created in the SZ when the rotating probe traversed along the center of the weld line. Although the probe crossed the weld line (at the center), no joint defects were found near and in the SZ for tool speeds ranging from 1,000 to 1,400 rpm. This result shows that the dissimilar FSW indeed has optimum welding conditions. The bonded interface is apparent in all cases, as crisscrosses near the SZ center. It is therefore accepted that there is no definite relationship between the speed of the tool and the structure of grains.

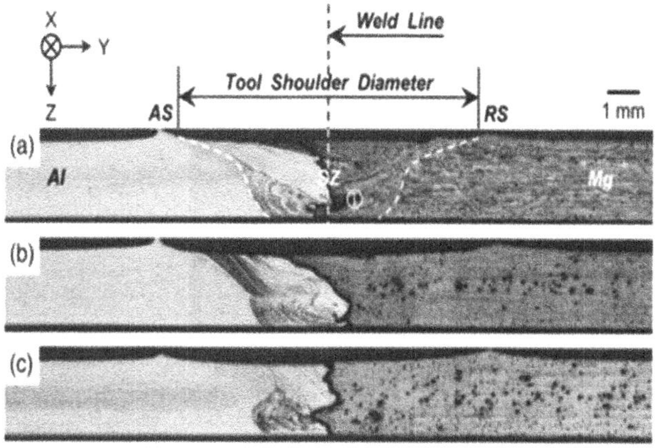

Figure 7.20. *Optical cross-section micrographs perpendicular to the friction-stir-welded plates along the traverse path of the tool with different rotational speeds, according to Y.J. Kwon* et al.

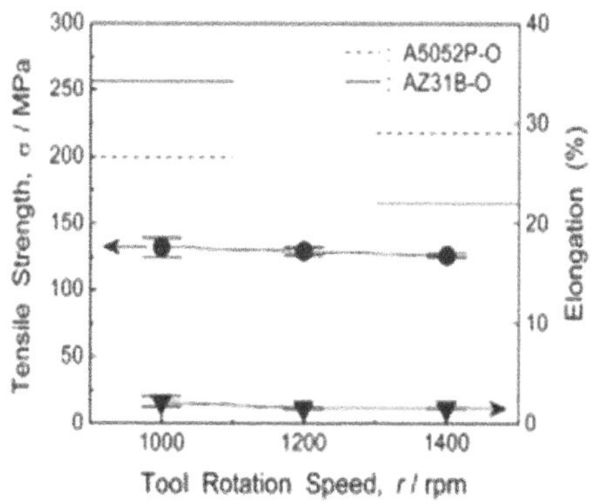

Figure 7.21. *Rotational speed of the tool influences the elongation and FSW plates. The split and solid lines represent the tensile strength and elongation of the base metals respectively, according to Y.J. Kwon* et al.

Figure 7.21 shows the impacts of the tool rotational speed on the material behavior of the FSWed sheets. The AZ magnesium 31B-O material has a higher tensile value (approximately 256 MPa) than the A5052P-O material (approximately 200 MPa). The maximum tensile value of about 132 MPa has resulted in FSW plates at the tool rotational speed of 1,000 rpm, which is about 66% of the tensile strength of the A5052P-O alloy. Nevertheless, the tensile value as a component of the speed of rotation (tool) did not change significantly. The FSWed sheets represented extraordinarily low deformation (2% or less). Elongation as an element of the tool rotational speed did not fundamentally change, although a maximum elongation of about 2% was obtained at 1,000 rpm.

Aruri Devaraju *et al.* (2019) conducted the microhardness test on a specimen at different tool rotational speeds (Figure 7.22). The increase of the rotational speed, on the one hand, improved the microhardness due to the adequate heat input delivered throughout the welding procedure, and, on the other hand, limited further growth of the grain. Moreover, effective and elevated stirring performance of the pin tool rotating results in small grains improving the microhardness.

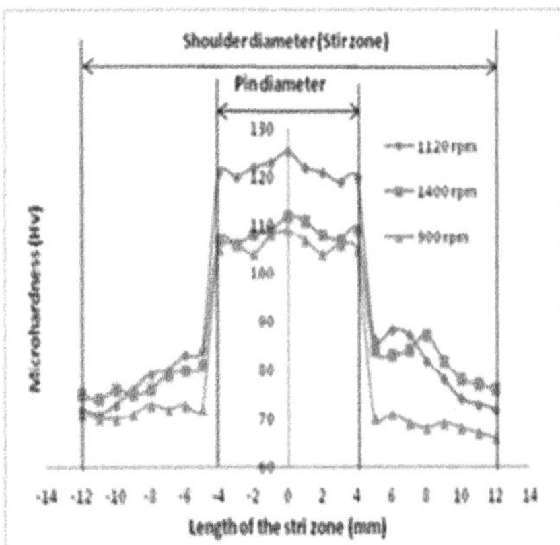

Figure 7.22. *Microhardness survey according to Aruri Devaraju* et al. *For a color version of this figure, see www.iste.co.uk/kumar/materials.zip*

The SEM translation of fractured samples reveals that differing types of disappointment regulate by small mixing of microvoid (Figure 7.23a–c). It shows that thermo-affected heat and welded chunk zones on two extraordinary pieces can be analyzed at three particular levels to see any imperfections of welding and the heat stream of the material. The material flow of heat is seen to be equivalently dispersed (Padmanaban and Balasubramanian 2009; Devaraju 2017).

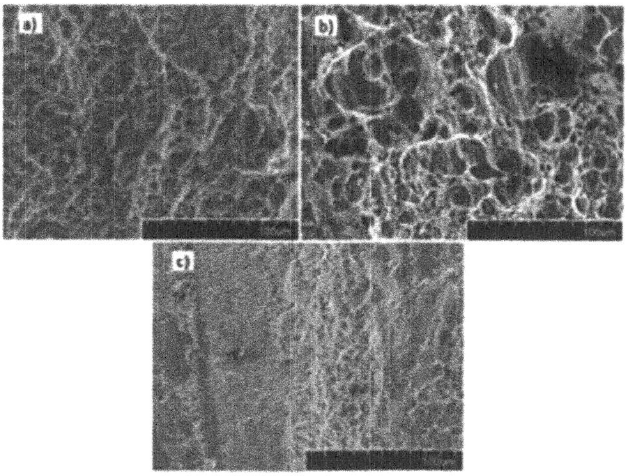

Figure 7.23. *Fracture features of dissimilar joints according to Aruri Devaraju* et al.

Sundaram *et al.* (2010) further showed that the tool's development in the material mass was smooth along the pin and plate surfaces. Likewise, when compared with the square tool, the rotating tool shoulder surface and the pin (round geometry) cause more plastic flow deformation on the grounds that the square tool's surface contact with the material mass is less (Suresha *et al.* 2011). Then, welded area bonding depends on the tool (rotational speed). For example, at lower speeds of 540 rpm, heat contribution to alloy plates is less, since it contains low (plasticized) stir zone. In addition, it brings the heat source towards the plates, causes inadequate flow of the material stir zone (plasticized condition) and neglects appropriate mixing of two alloys. However, at higher speeds, the material flow is faster and alloys are appropriately mixed, and subsequently the TS is higher (e.g. 1,400 rpm). The parameter welding speed contribute to the opposite effect on TS when related to the speed of the tool (in rotation) (Mohammad *et al.* 2015). The optimal limit parameters for the best TS are round pin dimensions, rotational speed of 1,400 rpm, welding speed 60 mm/min and tool (tilt angle) of 2°. The tests were aimed at those conditions and TS was assessed. It was observed that the variation between the evaluated TS and experimental estimates of TS was 6%. The micrographs were obtained at various procedural parameter variations and optimal process variables. All of the micrographs showed clear grain boundaries. One of these micrographs is shown in Figure 7.24 at optimal conditions. It affirms that when two alloys are mixed thoroughly, ideal grain limits can be achieved.

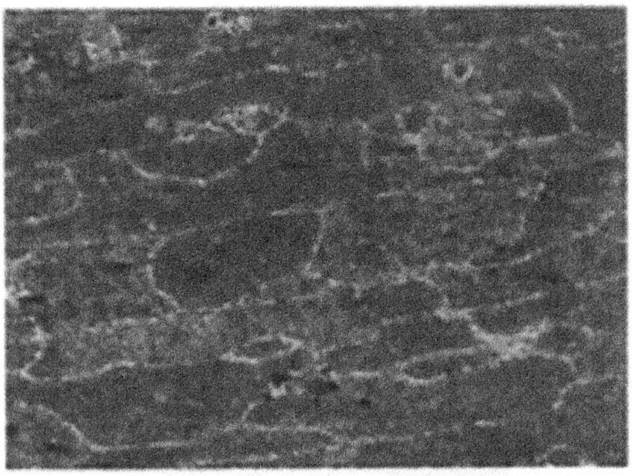

Figure 7.24. *Microstructure at optimal process parameters according to Shrinivas Balraj Uyyala* et al. *For a color version of this figure, see www.iste.co.uk/kumar/materials.zip*

H.M. Anil Kumar *et al.* (2020) showed the tensile test results of the R^2 estimate to be more than 80%, and therefore the model created was fit. The weld conducted using a shoulder diameter of 16 mm showed low tensile properties for the dissimilar aluminum alloys. The reason was attributed to the low frictional coefficient leading to less heat between the shoulder of the pin and the alloy. Furthermore, the weld was accomplished using a shoulder diameter greater than 20 mm, and almost more heat due to friction was created due to an increased overlap between the pin and alloy surface, albeit with low properties. The weld with an 18 mm pin shoulder diameter, 0 mm tool gap value and 1° tool tilt point indicated better tensile properties. Estimation of optimum FSW variables for a good weld leads to sufficient formability with fast recrystallization and recovery of grains within dissimilar materials. Figure 7.25 shows SEM images of sample no. 2. The parent metal interface zone AA 5083 and the weld nugget area are shown in Figure 7.25 (a). The fusion line is at the middle, and due to heat and stress the different zones of 5083 and 6082 result in expanded plasticity. The top surface of the nugget zone (NZ) with finely divided eutectic Mg_2Si particles form together the standard metals with the development of IMC (Figure 7.25(b)). The nugget's lower zone with divided elements of both the standard metal and substitute zone formation of two metals under upgraded softness is presented in Figure 7.25(c). In this case, a low extreme tensile resistance of 84.52 MPa is selected as a result of the chosen process parameters. Grain stream upward can be seen at AA6082 in the nugget zone in the thermo-mechanically affected zone (TMAZ), as shown in Figure 7.25(d). The EDAX micrograph (energy dispersive X-spectroscopy) range examination indicated a reduction in Mg grouping in the nugget layer because of the mixing of 5083 and 6082, where the elements were available from the two standard alloys as in the EDAX XRD spectrum shown in Figure 7.25(e).

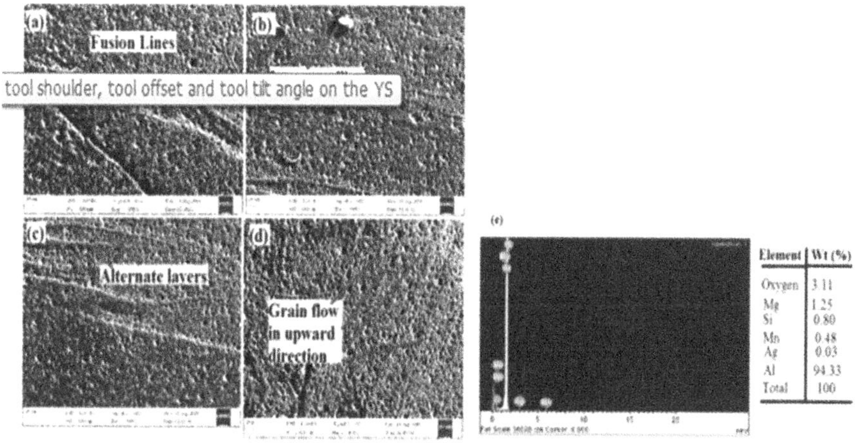

Figure 7.25. *SEM images of sample no. 2: (a) contact zone of AA5083, (b) top nugget zone, (c) bottom nugget zone, (d) TMAZ and (e) EDAX spectrum, according to H. Anil Kumar* et al. *For a color version of this figure, see www.iste.co.uk/kumar/materials.zip*

K. Palani *et al.* (2018) investigated the welded stir area microstructure using a hexagonal profile tool pin with a tool speed (rotational) of 1,500 rpm (Figure 7.26). Obviously, the dissimilar joint micrograph enormously relies on the tool inside the stir area. Micrographs witness movement of tools surface in contact with the alloy, improving friction and flow of material. In Figure 7.26(a) and (b), the microstructures of the standard metal alloys AA series 6061-T6 and AA8011 are shown. Equiaxed fine grains can be developed by selecting the best tool pin and the parameters of the FSW process, as shown in Figure 7.26(c). A greater speed (rotational) with a low weld speed creates more friction in the form of heat through plastic elongation and joint flow (material). In Figure 7.26(d), intermetallic compounds (IMC) are found to be brittle in nature at the AA6061-T6 aluminum alloy interface (Peel *et al.* 2003). Here, the hardness of the joints increases an adequate combination of the material with an adequate heat output where, in the aluminum alloy AA series 8011, quality diminishes due to joining hardening materials (cold condition) with minimal measurement of flow, as shown in Figure 7.26(e). The profile of the hexagon pin demonstrates better outputs in contrast to other joints (welded) (Taylor *et al.* 2008). The varieties of hardness are found because of the development of brittle intermetallic mixes (IMCs) at each side. In the 6061 aluminum alloy, the arrangement of intermetallic compounds (Al_3Mg_2) leads to a reduction in hardness in contrast to the hardness of the AA6061 aluminum compound.

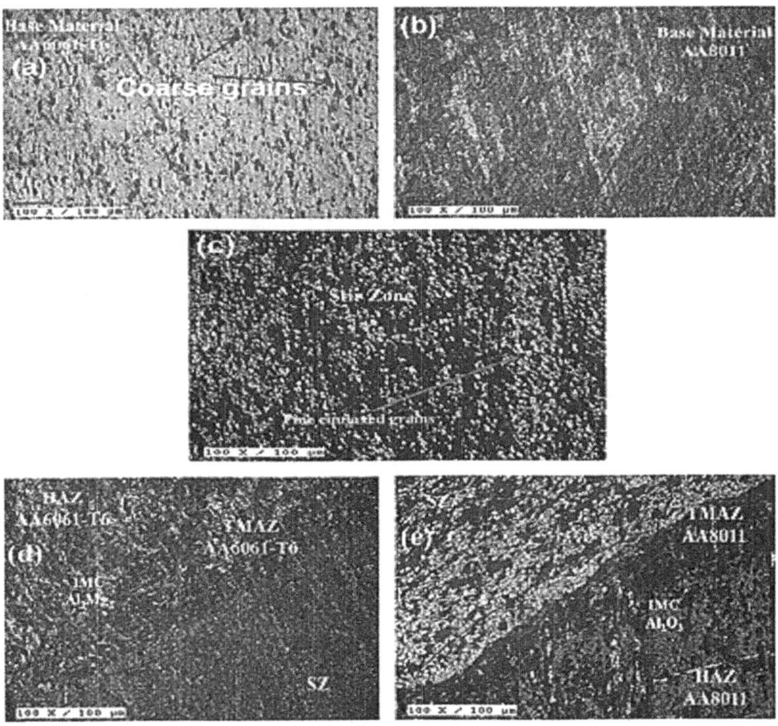

Figure 7.26. *Micrographs of weld using the hexagon tool pin profile of different alloys according to K. Palani* et al. *For a color version of this figure, see www.iste.co.uk/kumar/materials.zip*

A. Anitha Lakshmi *et al.* (2017) conducted experiments at strain rates of 0.01 s^{-1} to 0.0001 s^{-1} from room temperature to 700°C–900°C in intervals of 50°C in rolling [R0], perpendicular [R90] and transverse [R45] directions. Austenitic stainless steel (304) with a sheet thickness of 1 mm was described for testing tensile and formability. The study of the mechanical properties, anisotropy and strain rate sensitivity index of ASS 304 sheet materials at elevated temperatures showed that the mechanical properties of these rolled sheets were almost similar along different directions with r values very much close to 1.0. However, the rolled sheet showed high values in the transverse direction of rolling. This was because the ASS 304 material contained random distributions of strengthening precipitates. Anisotropy was studied regarding the effect of temperature. It was found that by initially increasing the temperature anisotropy from 700°C, then decreasing from 750°C to 800°C and increasing from 800°C to 850°C, a slight decrease follows up to 900°C. The average normal anisotropy was close to 1.0, which indicated good thinning resistance. The study of the effect of temperature on the sensitivity of strain rate (m) found that there was a considerable increase in the "m" value by increasing the temperature. Lade Jayahari *et al.* (2014) demonstrated that the austenitic stainless steel (ASS 304) of different blank diameters was drawn deep under warm conditions and the other material IS 737 grade aluminum alloy was drawn at elevated temperatures, and its formability was investigated under warm conditions. Moreover, 1 mm-thick circular blanks were made on an EDM wire-cutting machine of various diameters. The LDR calculated in the range of room temperature to 350°C represented a substantial LD ratio as the material temperature increased. The deep drawing experiments with similar conditions were done for the ASS 304 alloy. By tensile testing of materials, it was found that the highest value exceeded (cup) cracked slightly at the radius of punch. The deepest blank drawn was 65 mm by room temperature experiments, and the deepest blank dimension was 75 mm at 150°C and 300°C. Similar to Al, noteworthy enhancement of LDR at more temperatures similarly for ASS 304 (Hussaini *et al.* 2015; Kotkunde *et al.* 2017). A. Anitha Lakshmi *et al.* used ANN to predict the flow curve of AS steel 304, with the main aim of predicting the mechanical properties of AS steel 304 by uniaxial testing in an area created by super plastics. At temperatures ranging from 0 to 900°C, the tests were conducted at three variable stain rates at an interval of 50°C. The 90% tested data was used for training the ANN network and the remaining 10% tested data was used for checking the tested data. The average percentage error for the tested data was estimated to be 95% and the coefficient of correlation was 0.9 (Lakshmi *et al.* 2018a; Dharanikota *et al.* 2019; Lakshmi *et al.* 2019). A.A Lakshmi, *et al.* (2020) conducted the fractography analysis and constitutive modeling for dynamic plasticity of austenite stainless steel (ASS 304) at hot working temperatures S. Kosaraju *et al.* (2020) investigated the evaluation and characterization of ASS316L at sub-zero temperatures D. Baloji *et al.* (2020) compared ASS316L with formability at the room temperature and the super plastic region ul Haq *et al.* (2019) conducted the evaluation and optimization of material properties of ASS316L at elevated temperatures using the response surface methodology. K.V.D. Rajesh, *et al.* (2020) conducted the microstructure and corrosion resistance study on plasma arc welded joints of AISI 304 and AISI 316.

D. Mahesh Reddy *et al.* conducted tensile experiments at different strain conditions (0.1 s^{-1}, 0.001 s^{-1}) at 150°C and 450°C to determine the material behavior of Al8011 alloy sheets. The Taguchi orthogonal array and the (S/N) signal-to-noise ratio were used to evaluate the impact of considering chosen processes and their rates. The EDM tool was used for wire cutting. The Gray–Taguchi methodology was used for the optimization process. Taguchi's L8 array was chosen for experimental research based on the experimental results observed at low temperatures, and faster strain (0.1 s^{-1}) elongation rates were small compared to higher temperatures and lower strain levels (0.001 s^{-1}), taking into account all sheet thicknesses (Reddy *et al.* 2019). The amount of stress applied on the test decreased as the temperature increased, while higher or faster stress levels were needed for yield to begin. Slower strain levels indicated more elongation at either lower or higher temperatures. Test number 2 indicated higher grade of the gray connection and a lower S/N response curve. A. Anitha Lakshmi *et al.* developed experimental FLD by performing 800°C hemispheric dome punch tests on different specimens of duration. A theoretical FLD was developed using parameters for yielding from Marciniak–Kuczynski analysis with Hill 1948 and Barlat 1989. The theoretical FLD was compared with the experimental FLD, and confirmed. The substance used in this analysis was a 1.0 mm-thick layer of ASS 304. The yield locus showed that the yield estimate for the 1989 Barlat yield criteria included all the experimental data points. This assumed that Barlat's yield criteria for ASS 304 at 800°C in 1989 was suitable. In addition, the initial element of in homogeneity (thickness ratio) had a major effect on the estimation of limit formation. In this analysis, the rule of Hollomon force was used in conjunction with the parameters of yield of Hill 1948 and Barlat 1989. The result of changing the f0 value follows a general pattern. As the f0 value decreases, the FLD form remains the same, but in the negative y-direction the whole plot is displaced. The f0 value denotes the degree to which the groove region is thinner than the normal sheet area. This would make the groove more necking-prone as a result of an improvement in the rate of deformation compared with a smoother groove. This results in lower shape limits, which are defined by the FLD traversing in the negative y-direction. The f0 value for all of the yield parameters is calculated by the trial and error process and chosen as 0.995 for both (Lakshmi *et al.* 2018b). Theoretical FLD forecasts using yield criteria from Hill 1948 to underestimate the prediction of strain in the biaxial region (right-hand side). However, the FLD expected yield criterion of Barlat 1989 shows very close compliance with the findings of the experimental FLD.

7.4. Conclusion

In the present study, friction-stir welding was performed on AA6061 and AA2014 aluminum alloys by the LDH test. A non-heat treated TWB made up of AA6061 and AA2014 samples showed varied values of elongation, resulting in reduced formability. Solution treatment on the AA2014 alloy was made to obtain consistent elongation on each side of TWB. At this point, formability tests were carried out on the TWB, which also made use of the non-heat treated AA6061 alloy and a solution treated with the AA series 2014 alloy. A uniform tension test was performed on either of the TWBlanks on a section further treated by natural aging within 5 days.

In the present review, welding was performed on AA series 6061 and AA series 2014 aluminum alloys with various tool pin profiles using the friction-stir technology. Using the tensile and LDH tests, the crash pin profiles of tools on the mechanical properties and formability of welded blanks were explored. The blanks welded with square pin (tool profile) improved mechanical properties and formability when compared with blanks prepared with further pin profiles of tools. The energetic action of pin tool square in the weld zone created fine grains, which causes enhancement in weld area properties. The welded area microstructure of pin (square) reveals equally oriented grains and consistently spread grains. The use of the square pin profiled tool can make superior TWBs with aluminum alloys.

The 0.2% offset yield strength of the weld zone was twice as high as the yield quality of the sheets; however, the TS of the weld zone was about one and half times the parent sheet. Ductility and strain hardening of the weld area decreased, which was attributed to the higher weld zone quality. Normal anisotropy was found to be much lower than unity in the case of 1.4 mm-thick sheets, showing poor thinning resistance during the drawing process. While the average plastic strain ratio was found to be 0.94 in the case of the thicker sheet, suggesting that it was close to being isotropic. Experimental and numerical results showed that the height of the dome in the thicker sheet was higher than that of the thinner sheet, thus demonstrating the greater formability of the thicker sheet. The limiting dome height, due to TWB, was much less than the outcomes obtained for the parent sheets of dissimilar thicknesses, demonstrating a decreased TWB formability. A higher stress concentration in the thickness area in the weld line progressed towards a thicker cross-section. The strains increased as the dome center was approached, with the highest strain being near the necked area. The region where necking occurred lies matching the weld line and within the TWBs. The peak development of the weld line was at the center and equivalent to 2.57 mm.

Suitable feed and rotational speeds were 60 mm/min and 1,100 rpm respectively for completing the St14 FSW procedure. The stress–strain diagrams of these three zones revealed that the direction of the underlying grains had no effect on the welding properties (for tailored blanks of similar thickness). The proportion of variation in thickness also caused a decrease in strength.

The uniaxial tensile tests clearly showed that there were no significant differences between TWB's tensile strengths and those of its relative base metals.

In addition to extremely fine dispersed IMCs in aluminum, the FSW joint stirred zone showed an exceptionally rich matrix. The conveyance of hardness in the stirred zone was seen as deeply influenced by the consolidated impact of the fine grain size and the proximity of tough intermetallic compounds. The FSW test showed that there could be additional upgrades to unique welded aluminum and magnesium alloys.

In the nugget zone, more aluminum content was solvent compared to the AZ91 alloy where Mg was put on RS, affecting the material properties with respect to the micrographs. In the D1 specimen joint contact, where Mg was kept on the advancing side, shrill intermetallic zone was observed, where the dense intermetallic zone with weak aluminum magnesium attachment for the D2 specimen was observed.

Imperfection-free welds were successfully obtained at speeds (tool) of 1,000 to 1,400 rpm in increments of 200 rpm, recommending ideal speeds (tool) for various FSW. Additionally, SZ surface grain orientation converted smoothly with improvement in the speed (tool). The interface (simple bonded) was relatively clear when crisscrossing near the focal point of the SZ. There was no development of a eutectic microstructure, which suggested that, in the solid state of the base metals, the dissimilar FSW was carried out.

Increased Speeds (tool or machine) increase the microhardness due to the extra serious exciting use of pin rotation due to the higher heat input creation premise. Microstructure and fault observation were associated with the exploratory tensile properties and hardness.

The Taguchi technique was effectively applied to assess the impact on TS for welded joints of aluminum alloys AA series 6061-condition T6 and AA series 6082, of process parameters in the FSW method such as the tool tilt angle, the welding speed, the tool rotational speed and the pin geometry. The ANOVA tool was performed to distinguish the prevalent variables affecting the tensile strength. The pin geometry, the rotational speed and the welding speed of a tool are found to be influencing variables on the type of weld. From the Taguchi optimization technique, it is found that the welding speed of the tool is higher compared to the pin dimensions. The inclination of tilt tools was seen as a significant study for strong welding. Furthermore, it was assessed that TS was large in pin (circular) when compared with the square pin due to increasingly uniform material flow in the welding area. Microstructures showed clear grain boundaries at all the handling conditions, which affirmed the legitimate mix of the alloys.

Maximum tensile properties were recorded, such as ultimate 210 MPa elasticity, 203 MPa yield strength and 12% elongation using an 18 mm-diameter tool pin, with zero offset and 1° tool tilt. In addition, imperfection-free joints were supplied for high tensile strength samples. The microstructure examined the uncovering of suitable mix-up between the two standard metals and the scattering of huge measurements of Mg alloy particles in the nugget from the 5083 alloy.

Microstructural perception on dissimilar FSW blanks were examined, and imperfection-restricted (weld) joints with equally oriented grain structures were found at the stir area, accountable for more hardness and good joint strength (tensile). Intermetallic mix (Al3Mg2 alloy) lowered the hardness identified in the AA series 6061 aluminum alloys, while various IMCs (aluminum oxide) enhanced hardness compared with standard metal AA8011 alloys.

7.5. References

Adonyi, Y. and Chen, C. (1996). Laser-welded steel performance in tailored blank applications. *Automotive Automation Limited (England)*, 157–165.

Bhanodaya Kiran Babu, N., Davidson, M.J., Neelakanteswara Rao, A., Balasubramanian, K., and Govindaraju, M. (2014). Effect of differential heat treatment on the formability of aluminium tailor welded blanks. *Materials and Design*, 55, 35–42.

Bhanodaya Kiran Babu, N., Davidson, M.J., Neelakanteswara Rao, A., and Govindaraju, M. (2015). Formability analysis of dissimilar tailor welded blanks welded with different tool pin profiles. *Transactions of Nonferrous Metals Society of China*, 25, 1787–1793.

Chan, S.M., Chan, L.C., and Lee, T.C. (2003). Tailor-welded blanks of different thickness ratios effects on forming limit diagrams. *Journal of Materials Processing Technology*, 132, 95–101.

Chen, L.X., Bhandhubanyong, P., Vajragupta, W., and Somsiri, C. (1997). Plastic properties of low-carbon steel sheets. *Journal of Materials Processing Technology*, 63, 95–99.

Choi, Y., Heo, Y., Kim, H.Y., and Seo, D. (2000). Investigations of weld-line movements for the deep drawing process of tailor welded blanks. *Journal of Materials Processing Technology*, 108(1), 1–7.

Devaraju, A. (2017). Influence of post-weld rapid cooling on grain size and mechanical properties of friction stir welded AA2014. *Materials Today: Proceedings*, 4, 3722–3727.

Devaraju, A., Shalem, M.J., and Manichandra, B. (2019). Effect of rotation speed on tensile properties & microhardness of dissimilar Al alloys 6061-T6 to 2024-T6 welded via solid state joining technique. *Materials Today: Proceedings*, 18, 3286–3290.

Dharavath, B., ul Haq, A., Buddi, T., Singh, S.K., and Naik, M.T. (2020). Comparative study of ASS 316L on formability at room temperature and super plastic region. *Advances in Materials and Processing Technologies*, 6(2), 384–395.

Dilmec, M., Halkaci, H., Ozturk, F., Livatyali, H., and Osman, Y. (2013). Effects of sheet thickness and anisotropy on forming limit curves of AA2024-T4. *The International Journal of Advanced Manufacturing Technology*, 67, 2689–2700.

Ganesh Narayanan, R. and Naik, B.S. (2010). Assessing the validity of original and modified failure criteria to predict the forming limit of unwelded and tailor welded blanks with longitudinal weld. *Materials and Manufacturing Processes*, 25, 1351–1358.

Gautam, V. and Kumar, A. (2019). Experimental and numerical studies on formability of tailor welded blanks of high strength steel. *Procedia Manufacturing*, 29, 472–480.

Habibi, M., Hashemi, R., Fallah Tafti, M., and Assempour, A. (2018). Experimental investigation of mechanical properties, formability and forming limit diagrams for tailor-welded blanks produced by friction stir welding. *Journal of Manufacturing Processes*, 31, 310–323.

Haq, A.U., Kavit, A.K., Rao, T., Buddi, T., Baloji, D., Satyanarayana, K., and Singh, S.K. (2019). Evaluation and optimization of material properties of ASS 316L at elevated temperatures using response surface methodology. *Materials Today: Proceedings*, 18, 4589–4597.

Harshini, D., ul Haq, A., Buddi, T., Ajay Kumar, K., and Anitha Lakshmi, A. (2019). Comparative study on mechanical behavior of ASS 316L for low and high temperature applications. *Materials Today: Proceedings*, 19, 767–771.

Hussaini, S.M., Krishna, G., Gupta, A.K., and Singh, S.K. (2015). Development of experimental and theoretical forming limit diagrams for warm forming of austenitic stainless steel 316. *Journal of Manufacturing Processes*, 18, 151–158.

Jayahari, L., Sasidhar, P.V., Reddy, P.P., BaluNaik, B., Gupta, A.K., and Singh, S.K. (2014). Formability studies of ASS 304 and evaluation of friction for Al in deep drawing setup at elevated temperatures using LS-DYNA. *Journal of King Saud University – Engineering Sciences*, 26(1), 21–31.

Khan, A., Satya Suresh, V.V.N., and Regalla, S.P. (2014). Effect of thickness ratio on weld line displacement in deep drawing of aluminium steel tailor welded blanks. *Procedia Materials Science*, 6, 401–408.

Kosaraju, S., Kumar Singh, S., Buddi, T., Kalluri A., and ul Haq A. (2020). Evaluation and characterisation of ASS316L at sub-zero temperature. *Advances in Materials and Processing Technologies*, 6(2), 365–375. DOI: 10.1080/2374068X.2020.1728650.

Kotkunde, N., Krishna, G., Shenoy, S.K., Gupta, A.K., and Singh, S.K. (2017). Experimental and theoretical investigation of forming limit diagram for Ti-6Al-4 V alloy at warm condition. *International Journal of Material Forming*, 10(2), 255.

Kridli, G.T., Friedman, P.A., and Sherman, A.M. (2000). Formability of aluminium tailor-welded blanks. SAE Technical Paper, no. 2000-01-0772.

Kumar, H.A., and Ramana, V.V. (2020). Influence of tool parameters on the tensile properties of friction stir welded aluminium 5083 and 6082 alloys. *Materials Today: Proceedings*, 27(part 2), 951–957. Available at: https://doi.org/10.1016/j.matpr.2020.01.270.

Kumar, H.A., Raman, V.V., Shanmughanathan, S.P., John, J., and Iqbal, U.M. (2019). Optimization of dissimilar friction stir welding process parameters of AA5083-H111 and AA6082-T6 by CCD-RSM technique. In *Advances in Manufacturing Processes*, Vijay Sekar, K.S., Gupta, M., Arockiarajan, A. (eds). Springer, Singapore. Available at: https://doi.org/10.1007/978-981-13-1724-8_5.

Kusuda, H., Takasago, T., and Natsumi, F. (1997). Formability of tailored blanks. *Journal of Materials Processing Technology*, 71, 134–140.

Kwon, Y.J., Shigematsu, I., and Saito, N. (2008). Dissimilar friction stir welding between magnesium and aluminium alloys. *Materials Letters*, 62, 3827–3829.

Lakshmi, A.A., Rao, C.S., Gangadhar, J., Srinivasu, C., and Singh, S.K. (2017). Review of processing maps and development of qualitative processing maps. *Materials Today: Proceedings*, 4(2), 946–956.

Lakshmi, A.A., Rao, C.S., Srikanth, M., Faisal, K., Fayaz, K., and Singh, S.K. (2018a). Prediction of mechanical properties of ASS 304 in superplastic region using artificial neural networks. *Materials Today: Proceedings*, 5(2), 3704–3712.

Lakshmi, A.A., Rao, C., Kotkunde, N., Subbiah, R., and Singh, S.K. (2018b). Forming limit diagram of AISI 304 austenitic stainless steel at elevated temperature: Experimentation and modelling. *International Journal of Mechanical Engineering and Technology (IJMET)*, 9(12), 403–407.

Lakshmi, A.A., Srinivasa Rao, C., and Buddi, T. (2019). Prediction of superplasticity of austenitic stainless steel-304 at hot working temperatures. *Materials Today: Proceedings*, 18, 2814–2822.

Lakshmi, A.A., Buddi, T., and Srinivasa Rao, C. (2020). Fractography analysis and constitutive modeling for dynamic plasticity of austenite stainless steel (ASS 304) at hot-working temperatures. *Modern Manufacturing Processes*, 97.

Mohamed, M.A., Manurung Y.H.P., and Berhan, M.N. (2015). Model development for mechanical properties and weld quality class of friction stir welding using multi-objective Taguchi method and response surface methodology. *Journal of Mechanical Science and Technology*, 2323, 2331–2339.

Padmanaban, G. and Balasubramanian, V. (2009). Selection of FSW tool pin profile, shoulder diameter and material for joining AZ31B magnesium alloy – An experimental approach. *Materials and Design*, 30(7), 2647–2656.

Padmanabhan, R., Oliveira, M.C., and Menezes, L.F. (2008). Deep drawing of aluminium–steel tailor-welded blanks. *Materials & Design*, 29, 154–160.

Palani, K., Elanchezhian, C., Vijaya Ramnath, B., Bhaskar, G.B., and Naveen, E. (2018). Effect of pin profile and rotational speed on microstructure and tensile strength of dissimilar AA8011 and AA6061-T6 friction stir welded aluminium alloys. *Materials Today: Proceedings*, 5, 24515–24524.

Palanivel, R. and Koshy Mathews, P. (2011). The tensile behaviour of friction-stir welded dissimilar aluminium alloy. *Materiali in Tehnologije*, 45, 623–626.

Panda, S.K. and Ravi Kumar, D. (2009). Study of formability of tailor-welded blanks in plane-strain stretch forming. *International Journal of Advanced Manufacturing Technology*, 44, 675–685.

Paradiso, V., Rubino, F., Carlone, P., and Palazzo, G.S. (2017). Magnesium and aluminium alloys dissimilar joining by friction stir welding. *Procedia Engineering*, 183, 239–244.

Peel, M., Steuwer, A., Preuss, M., and Withers, P.J. (2003). Microstructure, mechanical properties and residual stresses as a function of welding speed in aluminium AA5083 friction stir welds. *Acta Materialia*, 51, 4791–4801. Available at: http://doi.org/10.1016/S1359-6454(03)00319-7.

Prasad, B.L., Neelaiah, G., Krishna, M.G., Ramana, S.V.V., Prakash, K.S., Sarika, G., Reddy, P.K., Dumpala, R., and Sunil, B.R. (2018). Joining of AZ91 Mg alloy and Al6063 alloy sheets by friction stir welding. *Journal of Magnesium and Alloys*, 6(1), 71–76.

Rajesh, K.D., Buddi, T., Kanth, P.R., and Satyanarayana, K. (2020). Microstructural and corrosion resistance study on plasma arc welded joints of AISI 304 and AISI 316. *Advances in Materials and Processing Technologies*, 6(2), 189–205.

Reddy, D.M., Lakshmi, A.A., and ul Haq, A. (2019). Experimental Taguchi approach and Gray–Taguchi optimization on mechanical properties of aluminium 8011 alloy sheet under uniaxial tensile loads. *Materials Today: Proceedings*, 19, 366–371.

Sameer, M.D. and Birru, A.K. (2019). Mechanical and metallurgical properties of friction stir welded dissimilar joints of AZ91 magnesium alloy and AA 6082-T6 aluminium alloy. *Journal of Magnesium and Alloys*, 7, 264–271.

Scriven, P.J., Brandon, J.A., and Williams, N.T. (1996). Influence of weld orientation on forming limit diagram of similar/dissimilar thickness laser welded joints. *Ironmaking and Steelmaking*, 23(2), 177–182.

Shanmuga Sundaram, N. and Murugan, N. (2010). Tensile behavior of dissimilar friction stir welded joints of aluminium alloys. *Materials & Design*, 31(9), Available at: 4184–4193.

Sharma, C., Dwivedi, D.K., and Kumar, P. (2012). Effect of welding parameters on microstructure and mechanical properties of friction stir welded joints of AA7039 aluminium alloy. *Materials and Design*, 36, 379–390. Available at: http://doi.org/10.1016/j.matdes.2011.10.054.

Sigeert, K. and Knabe, E. (1995). Fundamental research and draw die concepts for deep drawing of tailored blanks. SAE Technical Paper 950921, 866–876.

Suresha, C.N., Rajaprakash, B.M., and Upadhya, S. (2011). A study of the effect of tool pin profiles on tensile strength of welded joints produced using friction stir welding process. *Materials and Manufacturing Processes*, 26(9), 1111–1116.

Taylor, P., Elangovan, K., Balasubramanian, V., and Valliappan, M. (2008). Effect of tool pin profile and tool rotational speed on mechanical properties of friction stir welded AA6061 aluminium alloy. *Materials and Manufacturing Processes*, 23, 251–260. Available at: http://doi.org/10.1080/10426910701860723.

Yan, J., Xu, Z., Li, Z., Li, L., and Yang, S. (2005). Microstructure characteristics and performance of dissimilar welds between magnesium alloy and aluminum formed by friction stirring. *Scripta Materialia*, 53(5), 585–589.

8

Design and Analysis of Sedan Car B-pillar Outer Panel Using Abirbara with S-glass Fiber Hybrid Composites

Automotive industries are struggling to optimize vehicle bodies in different ways, such as shape for aerodynamics and aesthetics, weight of materials to be used for fuel efficiency, material conservation, recyclability and others. This century's competitive materials in the industry are composite materials, especially natural composites which are preferable from different perspectives. As the B-pillar is one of the critical structural support members of sedan cars, different considerations are used to analyze and replace its outer panel with natural nettle fiber composites. The B-pillar must overcome the stress developed due to the system as it is a structural member and safeguard the occupant in case of side crash. Considering these properties of the B-pillar, a natural fiber called Abirbara/Aleblabit/Ethiopian origin stinging nettle with an S-glass hybrid composite was used to replace its outer panel using Ansys analysis. It was observed that abirbara fibers, which can be extracted from the abirbara plant, are abundantly available in Ethiopia and are not yet used widely as composite materials. This work started with plant harvesting (collecting) and fiber extraction, which was performed through mechanical methods. The composites were prepared using chemically treated fiber with different orientations based on predetermined performances. Different mechanical properties were measured using UTM machines, such

Chapter written by Ramesh Babu NALLAMOTHU, Melkamu Yigrem YIHUNIE, Anantha Kamal NALLAMOTHU and Seshu Kishan NALLAMOTHU.

as tensile strength (153.166 MPa), compression strength (212.7 MPa) and bending strength (peak load 370 N), and water absorption was 1.6% in 48 hrs. The model of the sedan car B-pillar panel was developed using SpaceClaim software and analyzed for its impact using Ansys software. Crush simulation of the B-pillar panel were computed using conventional and composite materials. The Ansys analysis showed that the energy absorption of the EOSN fiber composite B-pillar panel was 2.61 kJ and that of conventional materials was 0.15925 kJ. It was concluded that abirbara composites can be used for the B-pillar outer panel, which reduces vehicle weight and fuel consumption as well as increases energy absorption.

8.1. Introduction

Automobile manufacturing industries use steel and its derivative materials to develop different structural and supportive components. As steel processing industries have become worldwide, many companies use steel rather than trying to investigate alternative materials.

The B-pillar not only houses electrical wiring and connection spots for passenger seat belts, but also provides structural support for the cab in case of a side collision or rollover of the vehicle. Vibrations generated from various sources should be considered in the design of a B-pillar. In this work, the outer panel of the B-pillar was designed using lightweight composite materials made of biodegradable natural fiber with a better energy absorption property.

Abirbara/Aleblabit/Ethiopian origin stinging nettle fiber is a plant of the species *Urtica dioica*, which is used as a source of fiber for domestic use (rope, whip) in the rural parts of Ethiopia, showing comparable mechanical properties (e.g. tensile strength, Young's modulus) with jute and other leaf-based fibers. This fiber has not yet been commonly used as a reinforcement in the preparation of composites. In the case of domestic applications, abirbara fiber has a comparable property with jute. It falls under the category of fibers extracted from the bast of the plant such as hemp, jute and flax.

The weight of materials in the automotive manufacturing sector influences fuel consumption, which is directly related to the corresponding amount of emission released into the atmosphere. As the weight increases, the consumption of fuel also increases, while the global fuel availability is running out besides the continuous cost increment. In addition to fuel constraints, Earths environmental pollution is also a major problem. Although there are different factors that contribute to air pollution, vehicle exhaust accounts for the maximum percentage. The weight of materials used can be considered as one of the factors for the high amount of emission released into the atmosphere when compared with what it takes due to the high amount of fuel consumption (Valerievich 2014). Natural fibers are abundant and renewable resources, so their cost is relatively low compared with other synthetic fibers. They are eco-friendly and bio-degradable, and reduce the problem of solid waste production by replacing non-degradable fillers. Natural fiber composites are relatively strong and lightweight materials, which can be used for the automobile body design and production instead of steels (Sanjay *et al.* 2016). Natural fibers from abirbara (Ethiopian origin stinging nettle), which have a higher amount of cellulose chemical components that are responsible for better mechanical properties, were selected in this work. Stinging nettle is classified into the shrub category with 30–45 species. It prefers fertile soil to grow to a height ranging from 40 to 120 cm. Stinging nettle is a natural biomass, which finds several applications in livestock, medicine, cosmetics and fibers. Nettle plants are covered with fine hair-like structures, especially in the leaves and stems. When touched, it releases chemicals that sting and trigger inflammation, causing redness, itching, bumps and irritation to the skin (Suryawan *et al.* 2017).

This work focuses on the development of composites, modeling and analysis of the B-pillar outer panel made from natural fiber composite materials to replace the existing steel. It reviews the literature related to sedan car body materials and their effects on human life, natural fibers and their composite material properties and composite manufacturing techniques. Ikpe (Ikpe *et al.* 2017) showed that lighter materials are developed to reduce the automobile's weight for cost and emission reduction. Among these lighter materials, composite materials were investigated for their probable use as impact energy absorbing elements. Some of the composites that have been investigated for use in crashworthiness are alloy metallic-reinforced composites. Alma Hodzic and Robert Shanks (2014) studied weight as one of the basic fuel economy factors for automobile design. Based on the test results, they concluded that fuel economy improvement for a weight decrement of 10% result was in the range of 6–7% fuel economy increment.

Ashokkumar and Beruk Hailu (2018) investigated the side impact of vehicles as this is the second most frequent type of accident, which risks passengers' lives. Composites have a better crash energy absorption property than the existing materials, although its use is not common in the industry, whereas processing the complex parts in one piece is much easier using composites. Moreover, the use of composite materials for the B-pillar reduces weight, which leads to less fuel consumption.

Wild stinging nettle, binomial name (*Urtica dioica*), is the most adaptable and abundant plant, which is found naturally in forests. Nettle fiber productivity showed 19–21% of the stem, of which 70% have primary long fibers (Rajasekar 2014). Ethiopian stinging nettle exhibits the following physical properties from a textile production perspective: length 5–75 cm, fineness 1.0–1.2 tex, diameter of the microscopic view at ×400 magnification 5–6 microns, moisture content 12.3% and dry tenacity 0.4–0.6 N/tex, along with breaking extension 2.0–2.5% (Tadele 2016). Fiber separation methods prove decreasing of non-cellulosic ingredients in the order of mechanical, water and chemical treatment methods. Kumaresan *et al.* (2015) concluded that composites have a significant economic potential when considering emerging advances in polymer composite body-in-white design against the mild-grade steel body currently on the road. With the significant implications of a polymer composite body for lightweight vehicles and therefore improved fuel efficiency, these results come at a time when they are particularly validated.

Natural fiber composites show environmentally better performance than glass fibers, while we use it independently due to the following reasons:

– natural fiber production has lower environmental impacts compared to synthetic fiber production;

– natural fiber composites have higher fiber content for equivalent performance, reducing more polluting base polymer content;

– lightweight natural fiber composites improve fuel efficiency and reduce emissions in the use phase of the component, especially in automotive applications;

– end-of-life incineration of natural fibers results in recovered energy and carbon credits (biodegradability).

Energy conservation and environmental pollution have become a serious concern around the world. European and many developed countries are researching and developing electric vehicles to reduce fuel consumption and exhaust emissions. They study how to increase the battery capacity to increase vehicle mileage. However, electric vehicles have an outstanding problem in use that the car is very heavy because of the battery weight, the charge capacity and the body structure, which in turn decreases the electric vehicle mileage (Ansari 2015). The B-pillar of a car is the structure located between the front and rear doors of the cab. According to Ansari's work, in the existing position of the B-pillar in a vehicle chassis, the upper and inner railings act as fixed members as they are fixed to the roof and bottom chassis of the vehicle. Thus, to carry out modal analysis on the B-pillar, we choose to treat the upper and lower railings as fixed supports. The weight of the composite B-pillar is reduced drastically when compared to the existing mild steel B-pillar. The gross weight of the composite B-pillar is estimated to be approximately 2.255 kg, whereas the weight of the present B-pillar made of steel is estimated to be 6.343 kg. The weight of the composite is 64.207% less than the steel B-pillar (Patil and Joglekar 2017). The important modification made by alkaline treatment is the disruption of hydrogen bonding in the network structure, thereby increasing surface roughness. The treatment changes the orientation of the highly packed crystalline cellulose order, forming an amorphous region. It has been reported that alkaline treatment has two effects on the fiber (Chubuike *et al.* 2017):

– it increases surface roughness, resulting in better mechanical interlocking;

– it increases the amount of cellulose exposed on the fiber surface.

Chemical treatments reduce some amount of hemicelluloses, lignin, pectin, wax and oil covering materials. Thus, the fiber surface becomes more homogeneous due to the elimination of microvoids. Stress transfer capacity between alternate cells improved and increased an effective fiber surface area for good adhesion with the matrix. It decreases the hydrophilic nature of the fiber by increasing its cellulose content and stabilizing the material. Natural fiber-reinforced polymer composites have beneficial properties such as low density, lower cost and reduced solidity when compared to synthetic composite products, thus providing advantages for use in commercial applications (automotive industry, buildings and constructions); Companies that are shifting to composites to design different components such as Ford (Mondeo CD 162, Focus) use composite for floor plates, door inserts, door panels, B-pillar and boot liner (Sardar 2014). Natural fibers have the potential to reduce vehicle weight. The use of natural fibers can minimize harmful pollutants, and their eventual breakdown is environmentally benign. When natural fibers break down, they emit less CO_2 than is absorbed during plant growth. Fiber-producing crops are easy to grow (Cioara 2016). Zinc borate shows excellent fire resistance performance and can therefore be widely used in polymer, rubber, fiber, paint and ceramic industries (Dholakiya 2009).

8.2. Materials and methods

8.2.1. *Materials*

The materials used in this work are discussed, starting from Ethiopian origin stinging nettle plant collecting up to the final B-pillar analysis. In this composite material work, two main composite components were considered: reinforcements and matrix constituents. Abirbara plant/EOSN and S-glass fibers (synthetic fibers), resin, wax, release agents, gel coat and miscellaneous items were used. The equipment used were a weighing machine, cloth (for swapping a wax-applied plat), stirrers, metal plates, weighting materials, a measuring cylinder and a universal testing machine. The materials used and equipment are shown in Figures 8.1–8.3.

Figure 8.1. *GP polyester resin. For a color version of this figure, see www.iste.co.uk/kumar/materials.zip*

Figure 8.2. *Maximum mold release wax. For a color version of this figure, see www.iste.co.uk/kumar/materials.zip*

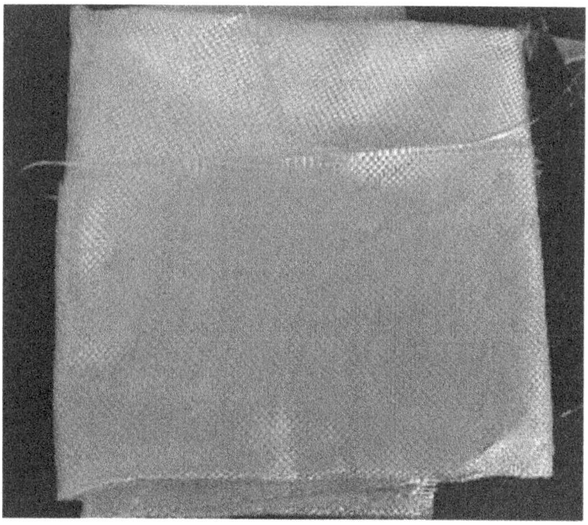

Figure 8.3. *S-glass. For a color version of this figure, see www.iste.co.uk/kumar/materials.zip*

UTM test machines were used in this work to conduct different mechanical properties of prepared composite materials, as shown in Figures 8.4–8.7.

Figure 8.4. *Tensile test. For a color version of this figure, see www.iste.co.uk/kumar/materials.zip*

Figure 8.5. *Impact tester. For a color version of this figure, see www.iste.co.uk/kumar/materials.zip*

Figure 8.6. *Bending test. For a color version of this figure, see www.iste.co.uk/kumar/materials.zip*

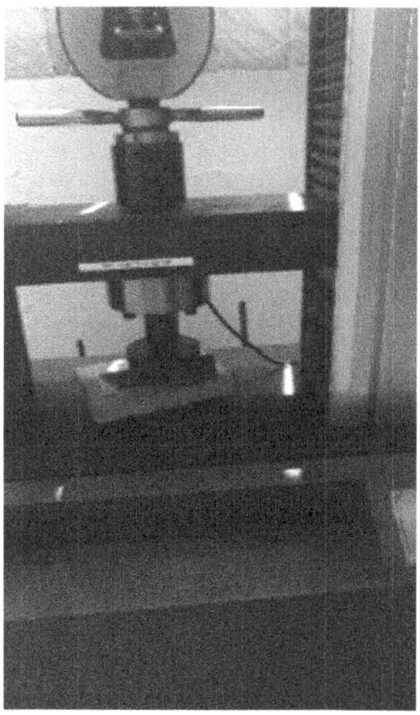

Figure 8.7. *Compression test. For a color version of this figure, see www.iste.co.uk/kumar/materials.zip*

8.2.2. *Methods*

Abirbara/Ethiopian origin stinging nettle was collected after drying in the field, which took approximately five to six months after germination. Its height ranged from 2 to 3 m depending on soil fertility. The plant harvesting process was performed using different equipment. In this work, traditional farm machinery such as a sickle (*machid*) and small axes were used. Figure 8.8 shows the collected Ethiopian origin stinging nettle.

Figure 8.8. *Ethiopian origin stinging nettle plant. For a color version of this figure, see www.iste.co.uk/kumar/materials.zip*

After collection, the fiber was extracted by two methods:

– In the first method, abirbara stems containing blast fibers were soaked in water for at least 10-12 hrs, then blast was removed from the stem and washed well for better quality. This process makes fiber extraction easy and protects the fiber from breakdown. However, during the process, some of the fibers were destroyed.

– In the second method, the stems were crushed first and then the blast was collected from the crushed stems as shown in Figure 8.9.

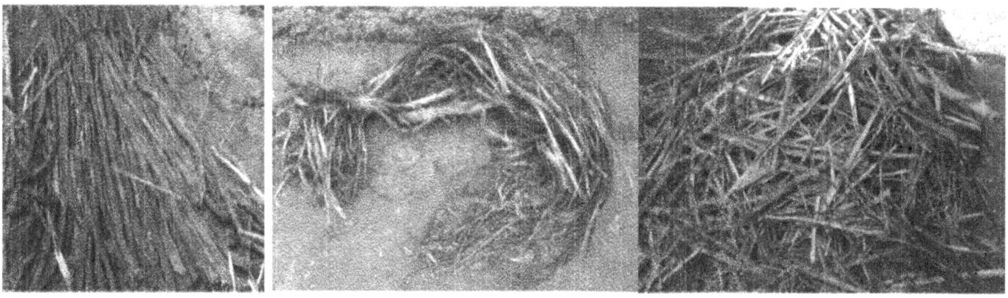

Figure 8.9. *Left to right: stem, abirbara fibers before processing and by-products. For a color version of this figure, see www.iste.co.uk/kumar/materials.zip*

The time required to soak abirbara fibers in NaOH depends on base concentrations that range from 2% to 10% in distilled water. Thus, the time course of physical changes occurring in the solution varied from 2 to 8 hrs, as shown in Figure 8.10. The fiber was washed with distilled and fresh water several times until the base/NaOH component was removed

Figure 8.10. *Soaking the fiber in the NaOH solution and drying. For a color version of this figure, see www.iste.co.uk/kumar/materials.zip*

8.3. Composite preparation, testing and analysis

8.3.1. *Composite preparation*

Fiber orientation determines the ability of composite to withstand against affective loads; for example, the fiber's ability to carry the load decreases as the angle between the load and the fiber direction increases. This leads to the separation of the polymeric matrix (general-purpose resin) and reinforcement fibers (abirbara fiber + S-glass fiber) as a result of the detached bonding between the matrix and fibers, where the forces stretch the matrix and cause the material to shear at the interface between the matrix and fibers, thereby leading to failure.

Figure 8.11. *0°/90° woven preparation. For a color version of this figure, see www.iste.co.uk/kumar/materials.zip*

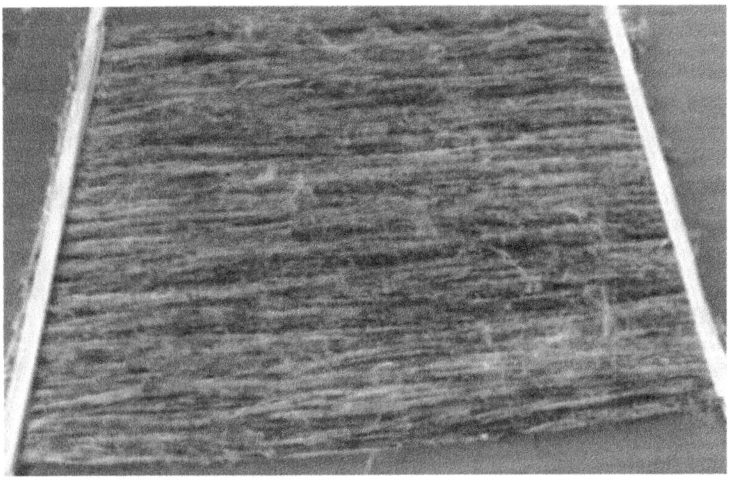

Figure 8.12. *0° and/or 90° fiber arrangement. For a color version of this figure, see www.iste.co.uk/kumar/materials.zip*

Figure 8.13. *$0^\circ/90^\circ$ woven arrangement. For a color version of this figure, see www.iste.co.uk/kumar/materials.zip*

Among the different processes used to develop FRC, the hand lay-up technique was used in this work by compromising its drawback, as shown in Figure 8.14. A square shape steel plate of size 410 mm was used for the hand lay-up process.

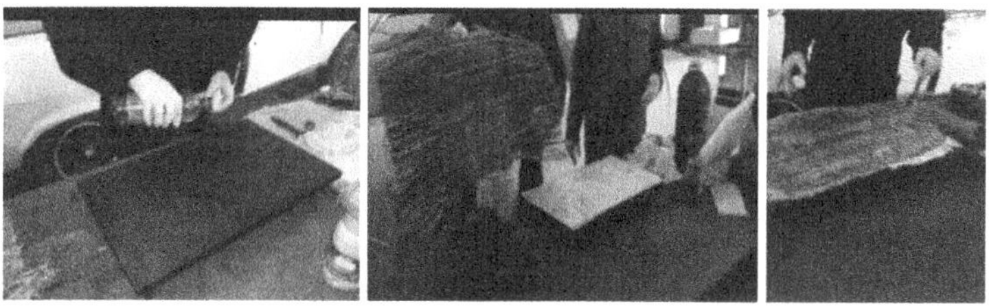

a) Applying the release agent b) Using EOSN fiber and applying

c) Composite plat

Figure 8.14. *Hand lay-up process. For a color version of this figure, see www.iste.co.uk/kumar/materials.zip*

Resin and fiber composition was determined by subtracting the summation of fiber weight from the composite weight. The nettle fiber weight was found to be 215.5 g and a single layer of S-glass woven was 68.0 g in one plat production. In this composite plat production with a size of 400 mm by

300 mm, two layers of glass fiber (one at the bottom and the other at the top) were used. Thus, the total mass of fibers used for this sample was 215.5 g + 2(68.0 g) = 351.5 g, while the matrix weight was found to be 865.5 g by subtracting from the plate weight which was 1217 g. Finally, the density was calculated as follows: $\rho = (1{,}217\ g/((40\times39.5\times0.545)\ cm^3)) = 1.4133\ g/cm^3$.

The composite preparation was generally carried out by the following steps:

– The purified and dried EOSN fiber was arranged according to the required fiber orientation; in this work, three fiber orientations with better performance (0°, 90° and 0°/90° woven) was used based on the literature.

– S-glass fiber woven was prepared for the size of the metal surface plate (mold).

– The weight of the nettle fiber and the S-glass fiber was determined.

– Polyester resin was mixed with 1% of catalyst and stirred for 15 min, in order to minimize the curing time of the composite.

– PVA and wax were applied on the mold surface using a clean cloth.

– First, a single S-glass layer was placed on the mold and then the matrix was applied, which dispersed well using brushes and a roller. After the resin matrix was spread well over the glass fiber, the EOSN fiber was arranged with a different orientation over the resin. Again, the matrix and the swab were applied carefully to avoid trapped air within the natural fiber, in order to prevent the formation of porosity that leads to crack initiation and subsequent failure. The third layer was added in the same manner as previously described.

– Finally, the load was applied to remove excess resin and to achieve a minimum thickness of the composite material.

8.3.2. *Testing and analysis*

Test specimens were prepared according to the ASTM standards using the band saw machine at Addis Ababa University, Ethiopia. The characterization of mechanical properties was according to the ASTM standards: tensile test according to ASTM D3039, compression test according to ASTM D695 and flexural test according to ASTM D6272. Each test was verified using an average of three samples for each of the three fiber orientations (0°, 90° and 0°/90° woven).

Figure 8.15 shows the average tensile strength values of the three types of fiber orientation after applying the correction factor. Hence, the B-pillar has a tensile strength of 153.166 MPa.

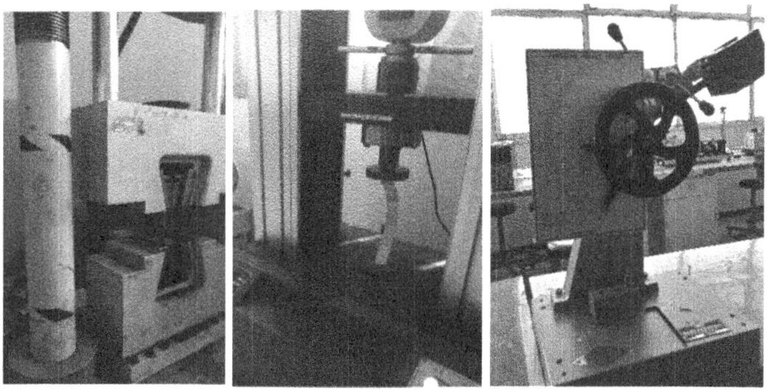

a) Tensile test b) Bending test c) Impact test

Figure 8.15. *Test machines. For a color version of this figure, see www.iste.co.uk/kumar/materials.zip*

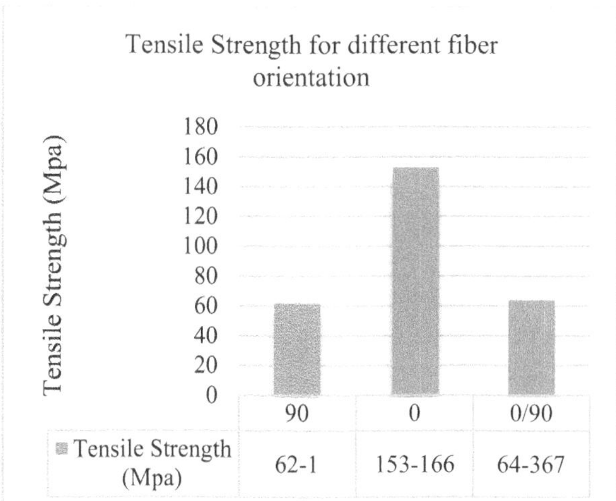

Figure 8.16. *Tensile strength of different fiber orientations. For a color version of this figure, see www.iste.co.uk/kumar/materials.zip*

Figure 8.16 shows the comparison of the peak load, deformation and compression strength for the different types of fiber orientation (0°, 90° and 0°/90° woven) obtained from the UTM machine display.

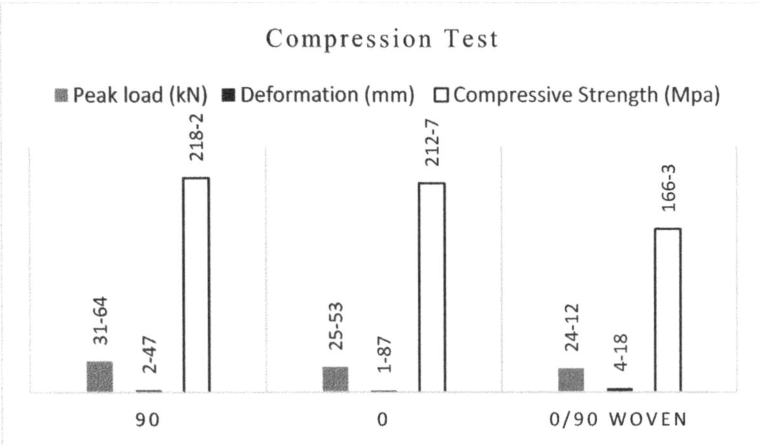

Figure 8.17. *Compression test results for different fiber orientations. For a color version of this figure, see www.iste.co.uk/kumar/materials.zip*

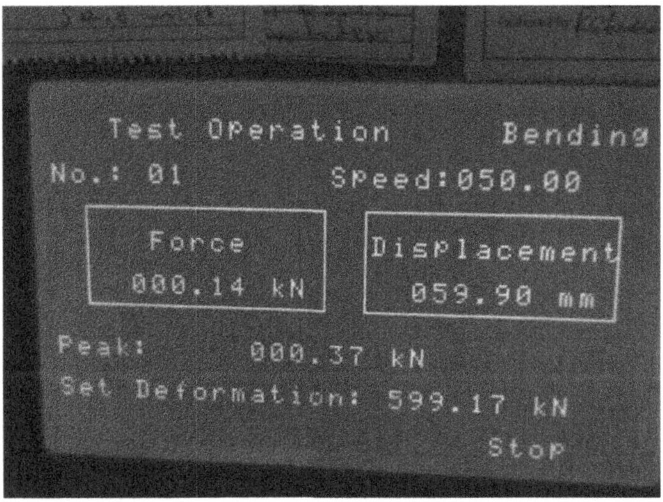

Figure 8.18. *0° fiber-oriented bending test result. For a color version of this figure, see www.iste.co.uk/kumar/materials.zip*

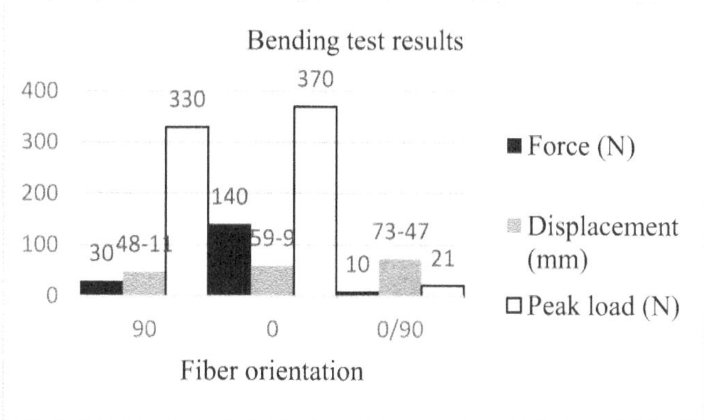

Figure 8.19. *Bending test comparison. For a color version of this figure, see www.iste.co.uk/kumar/materials.zip*

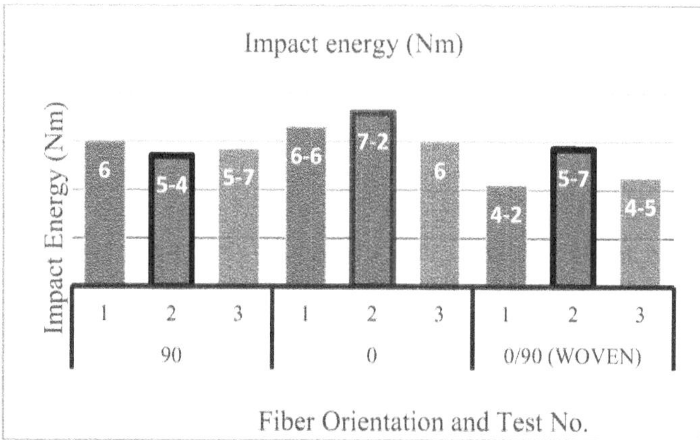

Figure 8.20. *Impact test comparison. For a color version of this figure, see www.iste.co.uk/kumar/materials.zip*

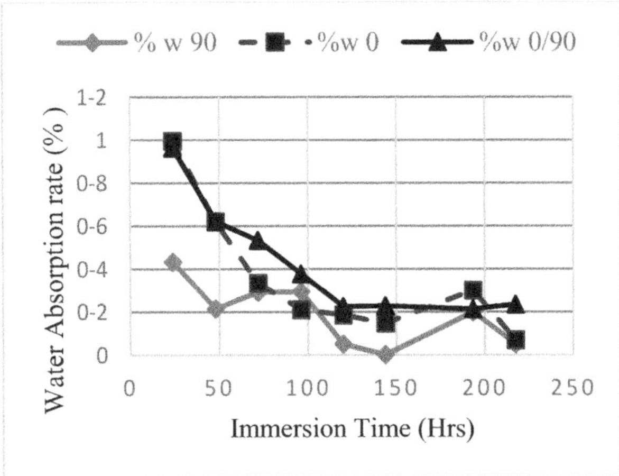

Figure 8.21. *Water absorption rate. For a color version of this figure, see www.iste.co.uk/kumar/materials.zip*

8.4. Design analysis of the B-pillar panel

Design analysis of the B-pillar outer panel of Toyota Corolla was modeled using Space Claim and was analyzed using Ansys R18.1, as shown in Figure 8.22.

In this work, we tried to compare the energy absorption property, weight reduction and others of the Ethiopian origin stinging nettle fiber and the S-glass fiber with a polyester resin matrix composite and carbon steel (which is the conventional material used for manufacturing the B-pillar).

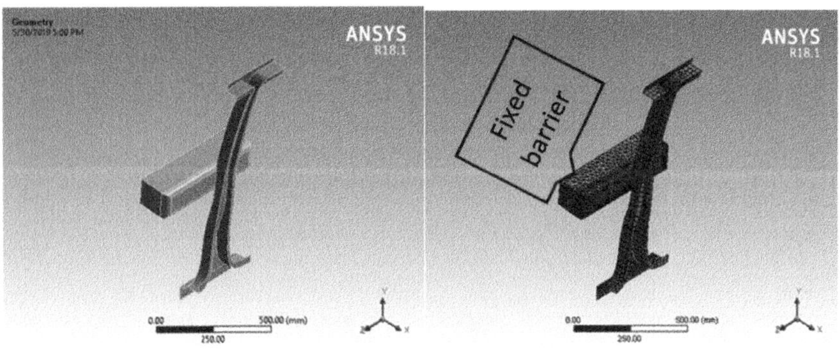

Figure 8.22. *Model for the B-pillar and the barrier element. For a color version of this figure, see www.iste.co.uk/kumar/materials.zip*

Material		
Assignment	Structural Steel	
Bounding Box		
Length X	161.63 mm	269.27 mm
Length Y	141.67 mm	1158.1 mm
Length Z	528.51 mm	380.54 mm
Properties		
Volume	1.0396e+007 mm³	3.9812e+005 mm³
Mass	81.608 kg	3.1252 kg
Centroid X	-12.749 mm	121.81 mm
Centroid Y	568.08 mm	540.57 mm
Centroid Z	-127.4 mm	-178.43 mm
Moment of Inertia Ip1	2.002e+006 kg·mm²	4.6953e+005 kg·mm²
Moment of Inertia Ip2	2.0437e+006 kg·mm²	14466 kg·mm²
Moment of Inertia Ip3	2.6914e+005 kg·mm²	4.6112e+005 kg·mm²
Surface Area(approx.)		3.3177e+005 mm²
Statistics		
Nodes	3193	1593
Elements	14803	1516
Mesh Metric	None	

Material		
Assignment	Structural Steel	EOSNF Composite
Bounding Box		
Length X	161.63 mm	269.27 mm
Length Y	141.67 mm	1158.1 mm
Length Z	528.51 mm	380.54 mm
Properties		
Volume	1.0396e+007 mm³	3.9812e+005 mm³
Mass	81.608 kg	0.56254 kg
Centroid X	-12.749 mm	121.81 mm
Centroid Y	568.08 mm	540.57 mm
Centroid Z	-127.4 mm	-178.43 mm
Moment of Inertia Ip1	2.002e+006 kg·mm²	84516 kg·mm²
Moment of Inertia Ip2	2.0437e+006 kg·mm²	2603.9 kg·mm²
Moment of Inertia Ip3	2.6914e+005 kg·mm²	83001 kg·mm²
Surface Area(approx.)		3.3177e+005 mm²
Statistics		
Nodes	3193	1593
Elements	14803	1517

Table 8.1. *Properties of the designed B-pillar outer panel for different materials*

	Mass (kg)	Volume (mm^3)	E (MPa)	$KE_{vehicle}$ (kJ)	σ_{ult} (MPa)	U_{max} (MPa) ($MN\text{-}m/m^3$)
EOSN panel	0.56254	3.9812e+5	1789.157	111.1	153.166	6.556
Steel panel	3.1252	3.9812e+5	200000	111.4	440	0.4

Table 8.2. *Comparison of the steel and EOSN B-pillar outer panels*

The total strain energy (toughness), U_{total}, that a B-pillar panel can absorb was determined by multiplying maximum strain energy by its volume for the composite material and steel respectively.

For the EOSN composite:

$$U_{total} = U_{max} \times \text{B-pillar volume}$$
$$= (6.556\text{MPa}) \times (0.00039812\text{m}^3)$$
$$U_{total,EOSN} = 0.00261\text{MN.m} = 2.61\text{KN.m}$$

Figure 8.23 shows the energy absorbed by the B pillar due to deformation during the collision.

The maximum von Mises stress of the side impact for the EOSN fiber composite B-pillar was 54.173 MPa. This stress developed in 4 msec, for a total of 0.02 sec. The total collision time and the stress distribution are shown in Figures 8.24 and 8.25.

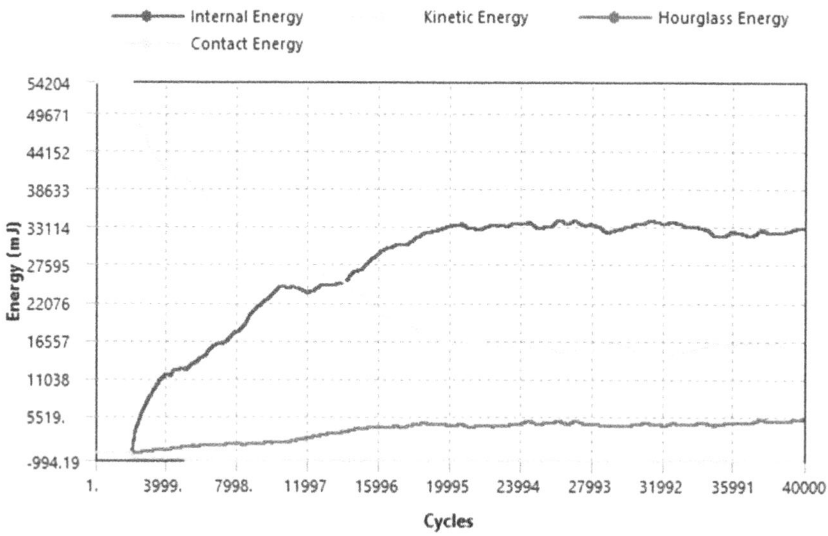

Figure 8.23. *Energy summary graph of the EOSN composite. For a color version of this figure, see www.iste.co.uk/kumar/materials.zip*

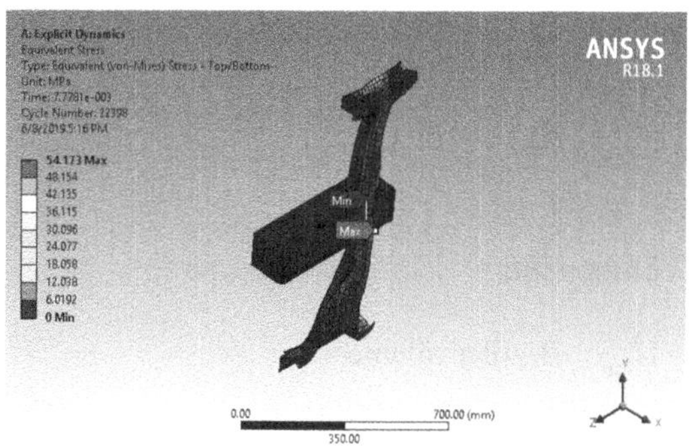

Figure 8.24. *Equivalent stress distribution of the EOSN composite. For a color version of this figure, see www.iste.co.uk/kumar/materials.zip*

Collision energy (KE) due only to its mass and specified speed was determined mathematically for AISI 1025 carbon steel B-pillar:

$$KE=\frac{1}{2}m_{steel\,panel}\times v^2$$
$$=0.5\times 3.1252kg\times(13.889m/s)^2$$
$$=301.43229J$$

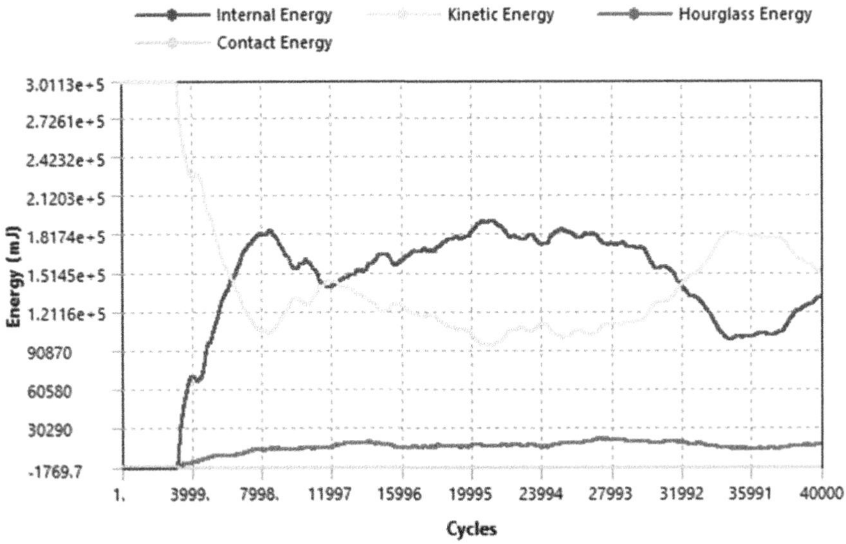

Figure 8.25. *Energy summary graph of the steel B-pillar outer panel. For a color version of this figure, see www.iste.co.uk/kumar/materials.zip*

For the steel component:

$$U_{total}=U_{max}\quad \text{B-pillar volume}$$
$$=(0.4MPa)\quad(0.00039812m^3)$$
$$U_{total,steel}=0.00015925MN.m=0.15925kN.m$$

Deformation variations between the composite and the steel B-pillar outer pillar were analyzed for the same boundary conditions. However, the maximum deformation of the composite and steel materials was 170.95 mm and 105.28 mm respectively.

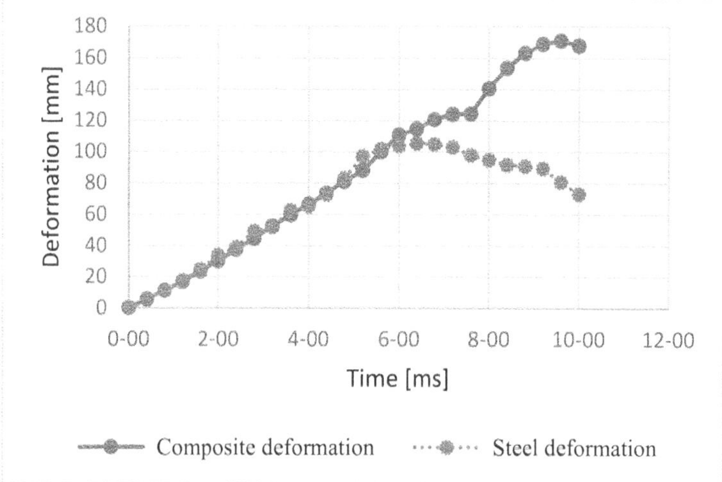

Figure 8.26. *Comparison of the deformation between the composite and the steel B-pillar outer panel. For a color version of this figure, see www.iste.co.uk/kumar/materials.zip*

In general, the EOSN composite newly designed B-pillar outer panel can absorb 2.61 kJ of energy out of vehicle colliding energy of 111.1 kJ, whereas the existing outer panel can absorb 0.15925 kJ of energy out of colliding energy of 111.4 kJ due to its speed and mass.

8.5. Conclusion

The B-pillar of the sedan car is a structural member with three components: internal panel, reinforcement (main part) and outer panel. The outer panel of the B-pillar was designed to be replaced with a natural fiber composite material to enhance vehicle performance.

Nettle is a plant that has more than 40 species around the world, which has not yet been studied in detail. Ethiopian origin stinging nettle contains fiber.

Three fiber orientations (0°, 90° and 0°/90°) were used for the nettle and S-glass layer with a polyester resin composite.

The 0° fiber-oriented composite had a tensile strength of 153.166 MPa, whereas the 90° fiber-oriented composite and 0°/90° woven had 62.1 MPa and 64.367 MPa respectively.

The 0° and 90° fiber-oriented composites had better compressive strength (212.7 MPa and 218.2 MPa respectively), whereas the 0°/90° woven had a compressive strength of 166.3 MPa. The bending test also showed that the 0° fiber-oriented composite deformed up to 59.9 mm at a peak load of 370 N.

The weight difference within 217.5 hr immersion in tap water was found to be 0.2244 g, 0.4801 g and 0.5729 g for 90°, 0° and 0°/90° woven respectively.

The abirbara composite with a fiber orientation of 0° was preferable to design the panel with an average tensile strength of 153.166 MPa and an average compression strength of 212.7 MPa. The software analysis was conducted based on these values. The energy absorption of the composite material B-pillar outer panel was 2.61 kJ of energy out of vehicle collision energy of 111.1 kJ, whereas the existing outer panel absorbed 0.15925 kJ of 111.4 kJ collision energy due to its speed and mass.

8.6. Recommendations

The age of Abirbara/stinging nettle plant has an effect on composite strength and the perfect time of cultivation after germination should be determined.

Using more accurate composite manufacturing techniques may increase output performances such as the resin transfer method, vacuum bagging and vacuum-assisted hand lay-up.

The EOSN natural fiber composite material property can be enhanced using a matrix with better strength, such as epoxy resin.

8.7. Acknowledgments

Above all, I would like to thank the Almighty God for His blessings in my life.

I would also like to express my special appreciation and gratitude to my advisor Dr. Ramesh Babu Nallamothu, Associate Professor for his help, guidance and support throughout this thesis. His continuous support was a source of motivation for me to undertake this research.

I would also like to thank all my friends for their kind and unreserved cooperation, encouragement and support during my thesis work.

Finally, I express my deep gratitude to my love and family for their unfailing support and continuous encouragement throughout the years of my studies and during my working on this thesis.

8.8. References

Ansari, M.N.M. (2015). A review on natural fiber reinforced polymer composite and its applications. *International Journal of Polymer Science*, 1–15.

Ashokkumar, T. and Hailu, B. (2018). A new accident proof material design for B – Pillar of a car. *International Journal of ChemTech Research*, 11(7), 1–11.

Chubuike, O.M., Ebele, C.C., Ifeanyi, I.F., Okwuchukwu, E.S., and Festus, O.E. (2017). Study on chemical treatments of jute fiber for application in natural fiber reinforced composites (NFRPC). *International Journal of Advanced Engineering Research and Science (IJAERS)*, 4(2), 21–26.

Cioara, L. (2016). Nettle fiber (urtica dioica), between myth and reality. Buletinul AGIR no. 3/2016, July–September, 19–25.

Dholakiya, B.Z. (2009). Use of non-traditional fillers to reduce flammability of polyester resin composites. Texas State University, San Marcos, USA.

Hodzic, A. and Shanks, R. (2014). *Natural Fiber Composites Materials, Processes and Properties*. Woodhead Publishing Limited, Cambridge, UK.

Ikpe, A.E., Owunna, I.B., and Satope, P. (2017). Design optimization of a B-pillar for crashworthiness of vehicle side impact. *Journal of Mechanical Engineering and Sciences*, 11(2), 2693–2710.

Kumaresan, M., Sathish, S., and Karthi, N. (2015). Effect of fiber orientation on mechanical properties of sisal fiber reinforced epoxy composites. *Journal of Applied Science and Engineering*, 18(3), 289–294. DOI: 10.6180/jase.2015.18.3.09.

Negera, D.W., Bhaskaran, J., Ilmi, I., and Nallamothu, R.B. (2019). Characterization of hybrid composite made of false banana fiber and sisal fiber. *International Journal of Engineering and Advanced Technology (IJEAT)*, 9(2), 3220–3226

Patil, B.D. and Joglekar, S. (2017). Vibration analysis of composite reinforced B-pillar used in automobiles. *International Journal of Innovative Research in Science, Engineering and Technology*, 6(7), 14242–14250. DOI: 10.15680/IJIRSET.2017.0607222.

Rajasekar, K. (2014). Experimental testing of natural composite material. *IOSR Journal of Mechanical and Civil Engineering (IOSR-JMCE)*, 11(2), 1–9.

Sanjay, M.R., Arpitha, G.R., Laxmana Naik, L., Gopalakrishna, K., and Yogesha, B. (2016). Applications of natural fibers and its composites: An overview. *Natural Resources*, 7(3), 108–114.

Sardar, K. (2014). Characterization and investigation of tensile test on kenaf fiber reinforced polyester composite material. *International Journal of Recent Development in Engineering and Technology*, 2(6), 104–112.

Suryawan, I., Suardana, N., Winaya, I.N., Suyasa, I., and Nindhia, T. (2017). Study of stinging nettle (urtica dioica l.) Fibers reinforced green composite materials. *7th International Conference on Key Engineering Materials (ICKEM 2017)*, 201(1), 1–7. DOI: 10.1088/1757-899X/201/1/012001.

Tadele, G.A. (2016). Separation and characterization of Ethiopian origin nettle fiber. *International Journal of Engineering Research & Technology (IJERT)*, 5(03), 259–262.

Valerievich, S.A. (2014). Advanced materials of automobile bodies in volume production. *European Transport/Trasporti Europei*, 56(10), 1–10.

List of Authors

Suresh AKELLA
Sreyas Institute of Engineering and Technology
Hyderabad
India

Zewdie ALEMAYEHU
Mechanical Engineering Department
Dilla University
Ethiopia

Ramesh Kumar BUDDU
FRMDC Division
Institute for Plasma Research
Bhat, Gandhinagar, Gujarat
India

Tanya BUDDI
Department of Mechanical Engineering
Gokaraju Rangaraju Institute of Engineering and Technology
Hyderabad,
Telangana
India

J. Paulo DAVIM
Department of Mechanical Engineering
Santiago Campus
University of Aveiro
Portugal

Dappu DEEPIKA
Department of Mechanical Engineering
Gokaraju Rangaraju Institute of Engineering and Technology
Hyderabad,
Telangana
India

Veeman DHINAKARAN
Center for Applied Research
Chennai Institute of Technology
India

Thimmaiah JAGADEESHA
Department of Mechanical Engineering
National Institute of Technology
Calicut
India

Jinesh Kumar JAIN
Department of Mechanical Engineering
MNIT
Jaipur
India

Hemalata JENA
School of Mechanical Engineering
KIIT University Bhubaneswar
Odisha
India

Jitendra Kumar KATIYAR
Department of Mechanical Engineering
SRM Institute of Science and Technology
Kattankulathur
Tamil Nadu
India

Kaushik KUMAR
Department of Mechanical Engineering
Birla Institute of Technology, Mesra
Ranchi
India

Akkireddy Anitha LAKSHMI
Department of Mechanical Engineering
Gokaraju Rangaraju Institute of Engineering and Technology
Hyderabad,
Telangana
India

Mekonnen LIBEN
Mechanical Systems and Vehicle Engineering Department
SoMCME
Adama Science and Technology University
Ethiopia

Jayakumar MANI
Haramaya University
Dire Dawa
Ethiopia

Anantha Kamal NALLAMOTHU
Department of Mechanical Engineering
Politecnico Di Milano
Lecco
Italy

Ramesh Babu NALLAMOTHU
Mechanical Systems and Vehicle Engineering Department
SoMCME
Adama Science and Technology University
Ethiopia

Seshu Kishan NALLAMOTHU
Department of Mechanical Engineering
Politecnico Di Milano
Bovisa
Italy

Muthuvelayudham RAMANUJAM
Annamalai University
Tamil Nadu
India

Chalamalasetti Srinivas RAO
Department of Mechanical Engineering
Andhra University
Vishakapatnam, Andhra Pradesh
India

Pankaj SONIA
Department of Mechanical Engineering
MNIT
Jaipur
India

Bathini SRIDHAR BABU
Department of Mechanical Engineering
CMR Institute of Technology, Hyderabad
Telangana
India

Madabushi SWAPNA SAI
Center for Applied Research
Chennai Institute of Technology
India

Selvaganapathy THAMBIYAPILLAI
Tamilnadu Pollution Control Board
Tamilnadu
India

Mahesh VARSHA SHREE
Center for Applied Research
Chennai Institute of Technology
India

Harinadh VEMANABOINA
Sri Venkateswara College of Engineering and Technology
Chittoor, Andhra Pradesh
India

Melkamu Yigrem YIHUNIE
Debre Tabor University
Ethiopia

Index

A, B, C

D, E, F

G, L, M, P

Q, R, S, T

Other titles from

iSTE

in

Materials Science

2019

SALENÇON Jean
Viscoelastic Modeling for Structural Analysis

2018

ABABOU Rachid
Capillary Flows in Heterogeneous and Random Porous Media 1

GODIN Nathalie, REYNAUD Pascal, FANTOZZI Gilbert
Acoustic Emission and Durability of Composite Materials
(Durability and Ageing of Organic Composite Materials Set – Volume 3)

JANNOT Yves, DEGIOVANNI Alain
Thermal Properties Measurement of Materials

2017

CAILLETAUD Georges, SAÏ Kacem, TALEB Lakhdar
Multi-mechanism Modeling of Inelastic Material Behavior

CARVELLI Valter, JAIN Atul, LOMOV Stepan
Fatigue of Textile and Short Fiber Reinforced Composites
(Durability and Ageing of Organic Composite Materials Set – Volume 1)

DELHAES Pierre
Materials and Thermodynamics: Living and Economic Systems

FRESSENGEAS Claude
Mechanics of Dislocation Fields

GIGLIOTTI Marco, LAFARIE-FRENOT Marie-Christine, GRANDIDIER Jean-Claude, MINERVINO Matteo
Mechanical Behavior of Organic Matrix Composites
(Durability and Ageing of Organic Composite Materials Set – Volume 2)

2016

TORRENTI Jean-Michel
Materials and Infrastructures 1
Materials and Infrastructures 2

RÉMOND Yves, AHZI Saïd, BANIASSADI Majid, GARMESTANI Hamid
Applied RVE Reconstruction and Homogenization of Heterogeneous Materials

2015

JOUD Jean-Charles, BARTHÉS-LABROUSSE Marie-Geneviève
Physical Chemistry and Acid-Base Properties of Surfaces

LALEVÉE Jacques, FOUASSIER Jean-Pierre
Dyes and Chromophores in Polymer Science

2014

HAMAIDE Thierry, DETERRE Rémi, FELLER Jean-François
Environmental Impact of Polymers

LAFLEUR Pierre, VERGNES Bruno
Polymer Extrusion

2013

BATHIAS Claude
Fatigue Limit in Metals

LEXCELLENT Christian
Shape-memory Alloys Handbook

2012

DAVIM J. Paulo
Wear of Advanced Materials

DELHAES Pierre
Carbon Science and Technology: From Energy to Materials

GOUDEAU Philippe, GUINEBRETIÈRE René
X-Rays and Materials

SAANOUNI Khemais
Damage Mechanics in Metal Forming: Advanced Modeling and Numerical Simulation

VANBÉSIEN Olivier
Artificial Materials

VERDU Jacques
Oxidative Ageing of Polymers

2011

BATHIAS Claude, PINEAU André
Fatigue of Materials and Structures: Application to Damage

BATHIAS Claude, PINEAU André
Fatigue of Materials and Structures: Application to Design

PRIESTER Louisette
Grain Boundaries and Crystalline Plasticity

VIGNES Alain
Extractive Metallurgy 1: Basic Thermodynamics and Kinetics
Extractive Metallurgy 2: Metallurgical Reaction Processes
Extractive Metallurgy 3: Processing Operations and Routes

2010

BATHIAS Claude, PINEAU André
Fatigue of Materials and Structures: Fundamentals

CHATEIGNER Daniel
Combined Analysis

CHEVALIER Yvon, VINH TUONG Jean
Mechanical Characterization of Materials and Wave Dispersion V1
Mechanical Characterization of Materials and Wave Dispersion V2

DELHAES Pierre
Carbon-based Solids and Materials

2009

ALVAREZ-ARMAS Iris, DEGALLAIX-MOREUIL Suzanne
Duplex Stainless Steels

DAVIM J. Paulo
Machining Composite Materials

GALERIE Alain
Vapor Surface Treatments

2008

BATTAGLIA Jean-Luc
Heat Transfer in Materials Forming Processes

BLONDEAU Régis
Metallurgy and Mechanics of Welding

FRANÇOIS Dominique
Structural Components

REYNE Maurice
Plastic Forming Processes

TAKADOUM Jamal
Materials and Surface Engineering in Tribology

2007

BOCH Philippe, NIEPCE Jean-Claude
Ceramic Materials

CRISTESCU Constantin
Materials with Rheological Properties

2006

LASSEN Tom, RÉCHO Naman
Fatigue Life Analyses of Welded Structures